Fab Lab

Revolution Field Manual

Fab Lab – Revolution Field Manual

The Deutsche Nationalbibliothek
lists this publication in the Deutsche
Nationalbibliografie; detailed
bibliographic data are available on
the Internet at http://dnb.dnb.de

ISBN 978-3-7212-0965-5

© English edition 2017 Niggli, imprint
of Braun Publishing AG, Salenstein
www.niggli.ch

A Vetro Editions project
www.vetroeditions.com

Editor: Massimo Menichinelli

Authors: Massimo Menichinelli,
Camille Bosqué, Peter Troxler,
Cecilia Raspanti, Alex Schaub,
Heloisa Neves

Cover design: adaptation of
the lettering for Labfab
www.labfab.fr – Benjamin Grosfilley – BUG

Design: Luca Bendandi – Vetro Editions

Copyediting: John Z. Komurki

Infographics: Matteo Astolfi

Fab Lab

Revolution Field Manual

Edited by
Massimo Menichinelli

Written by
Massimo Menichinelli
Camille Bosqué
Peter Troxler
Cecilia Raspanti
Alex Schaub
Heloisa Neves

niggli

The Fab Charter

What is a Fab Lab

Fab Labs are a global network of local labs, enabling invention by providing access to tools for digital fabrication

What is in a Fab Lab

Fab Labs share an evolving inventory of core capabilities to make almost anything, allowing people and projects to be shared

What does the Fab Lab network provide

Operational, educational, technical, financial, and logistical assistance beyond what is available within one lab

Who can use a Fab Lab

Fab Labs are available as a community resource, offering open access for individuals as well as scheduled access for programs

What are your responsibilities

Safety: not hurting people or machines

Operations: assisting with cleaning, maintaining, and improving the lab

Knowledge: contributing to documentation and instruction

Who owns Fab Lab inventions

Designs and processes developed in fab labs can be protected and sold however an inventor chooses, but should remain available for individuals to use and learn from

How can businesses use a Fab Lab

Commercial activities can be prototyped and incubated in a fab lab, but they must not conflict with other uses. They should grow beyond rather than within the lab, and they are expected to benefit the inventors, labs, and networks that contribute to their success

DRAFT: OCTOBER 20, 2012

HTTP://FAB.CBA.MIT.EDU/ABOUT/CHARTER/

Neil Gershenfeld's celebrated, visionary Fab Lab at the Massachusetts Institute of Technology enables anybody to design and execute one-of-a-kind objects complete with brains.

The New York Times

These labs form part of a larger Maker movement of high-tech do-it-yourselfers, who are democratizing access to the modern means to make things.

Neil Gershenfeld

4 5 6 7

Massimo Menichinelli

Introduction

The story of the Fab Lab network is the story of how innovations can sometimes start and spread by accident. It is also the story of how globalization can give rise to collaborative networks between countries, rather than exclusively top-down supply chain networks. More than fifteen years have passed since the launch of the Center for Bits and Atoms at MIT, where everything started with the first Fab Lab. Led by Prof. Neil Gershenfeld, the Center launched the first lab not only for research but also as an example of how digital fabrication could affect our lives.

Following this example, many more labs were established in many countries, at the beginning in imitation of the first lab, and then as the result of more elaborate and independent initiatives. Since then, the network of labs has grown almost exponentially to more than 1000 Fab Labs at the time of writing of this book. We are now witnessing the end of the first phase in the history of Fab Labs, where most effort went on further refining the concept and launching new labs. These processes will still be important, but the current frontier that needs to be explored by the network is how to make such labs financially, socially and environmentally sustainable, and how to connect the labs with for-profit companies, for example.

But what is a Fab Lab? It is a workshop for making physical objects with digital fabrication technologies, often in a collaborative way. But not every workshop is a Fab Lab: there are four different conditions that a lab needs to meet in order to be able to call itself a Fab Lab. First, the lab must provide public access, even if it is limited. Second, the lab must support and subscribe to the Fab Charter, the Fab Lab manifesto. Third, the lab must have the same set of tools and processes as all the other labs, in terms of typologies (i.e. not necessarily brand-specific): these are mostly digital fabrication technologies, but handwork processes are also included. Fourth, the lab must collaborate with the other labs in the Fab Lab network: an isolated lab is not a Fab Lab.

These are the common requirements for Fab Labs, but there are many more details that characterize them. Since each Fab Lab needs to adapt the global model to its local context, there are many local variations and more specific features. Furthermore, there are different perspectives and philosophies regarding the concept of Fab Labs, how to run them, and how to coordinate the global network. The diversity in the Fab Lab network is what is making it so successful at the moment, and will ensure its sustainability in the long term. The importance of multiple perspectives on the Fab Lab community is the reason more than one author was invited to participate in this book, and interviews included: this is an international book, which we hope will reflect the different ideas afloat in the international network.

In **Chapter One**, Camille Bosqué tells us the story of the Fab Lab network through the history of the first Fab Labs, which she has visited and investigated as part of her research. In this chapter we discover how the network sprung up from a single lab at MIT.

In **Chapter Two**, Massimo Menichinelli describes the most common technologies and processes available in Fab Labs. Most of them are digital fabrication technologies, but there are also analog machines and handwork-based processes. Not all digital fabrication technologies can be found in Fab Labs, for many reasons: they may be too expensive or too difficult to operate, and it would be difficult successfully to democratize access to them. These technologies may find their way into Fab Labs in the future, but at the moment they are not available and are not representative of what can be manufactured in Fab Labs. At the same time, all of these technologies and processes can be, and most of the time are, combined: molding and casting is a process that can be done after CNC milling a countermold, for example.

In **Chapter Three**, Peter Troxler critically addresses the philosophy behind Fab Labs, their purpose and application and the different kinds of Fab Labs that can be generated. The chapter ends with a reflection on the diverse challenges that the Fab Lab community will have to address in the future, in order to have a real impact and become sustainable, and avoid the bubble effect.

In Chapter Four, Massimo Menichinelli explores the business dimension of Fab Labs, analyzing the current emerging economy of digital fabrication, the process for developing a Fab Lab and its business model and business plan, and the most common business models and approaches for the projects developed in Fab Labs. The chapter includes the profiles of three different labs that Massimo Menichinelli has developed, to show how each lab was developed in a different way and with a different business model for each specific context.

In **Chapter Five**, Cecilia Raspanti and Alex Schaub give us a great example of how projects can be developed in Fab Labs, showcasing the process and the results of Cecilia Raspanti's internship at Fablab Amsterdam. There were several projects developed in this internship, all of them exploring a different side of how fashion design could be transformed by the technologies and processes available in Fab Labs, from the development of custom objects to the development of custom tools.

In **Chapter Six**, Massimo Menichinelli looks at some representative projects that illustrate what can be achieved in Fab Labs at the moment. Some of the projects were designed and manufactured in Fab Labs, other were designed outside Fab Labs and then manufactured within them, while some were designed and manufactured outside of Fab Labs but could be easily replicated there, because they are based on the technologies and approaches applied by the Fab Lab network. There are different kinds of objects, technologies and approaches in these projects - some of them are prototypes, others are final products; some of them are shared as open source projects, others are proprietary and closed projects. Most of these projects are the outcome of a collaboration between different people. As a whole, they show us how design and manufacturing can change thanks to Fab Labs, what their limits are, and what is actually done there.

In **Chapter Seven**, Heloisa Neves tells the story of some of the pioneers in the global Fab Lab network whom she has met in her research. She also interviewed some of them in depth, showcasing the different experiences and approaches that such pioneers have developed in different continents.

Finally, **Chapter Eight** covers further resources related to the Fab Lab network.

Neil Gershenfeld started promoting the concept of Fab Labs with FAB, the first book about Fab Labs, in 2005. Since then, very few books on the topic have been published: with this one, we are trying to cover the history so far and the challenges ahead, while at the same time setting out what comprises a Fab Lab, and what can be achieved inside one.

Camille Bosqué

History of Fab Labs

Origins, purposes, development and new directions

Today, Fab Labs can be found all over the planet, but when the first lab was launched there was no plan to establish a worldwide network. Following its inauguration, more and more communities started to become interested in the project, and the first Fab Labs outside MIT were created, refining and diffusing its principles. The concept has become widespread and the network of labs has grown proportionally, with new labs being created almost every day. This chapter covers the history of the first Fab Labs that were set up beyond MIT, transforming an experiment into a blossoming global network.

Milestones in the history of Fab Labs

The Fab Foundation was formed in order to facilitate and support the growth of the international Fab Lab network through the development of regional Fab Foundations and organizations

Fifth Fab Lab meeting in Pune (India): FAB5: The Fifth International Fab Lab Forum and Symposium on Digital Fabrication[7]

The Fab Academy course, organized among many Fab Labs worldwide, is launched

2009

2008

Launch of Fab Lab Vestmannaeyjar in Iceland.[6]

First Fab Lab meeting at MIT: Fab Lab User Group Meeting[1]

Second Fab Lab meeting in Ørnes, Lyngseidet (Norway): Symposium on Digital Fabrication and Norwegian Fab Lab Opening[2]

First book about Fab Labs: Gershenfeld, N., 2005. FAB: The Coming Revolution on Your Desktop – From Personal Computers to Personal Fabrication, Basic Books.

2005

2004

First Fab Lab in Africa at Takoradi Technical institute (Ghana)

Fourth Fab Lab meeting in Chicago (USA): FAB 4: The Fourth International Fab Lab Forum and Symposium on Digital Fabrication[5]

Fab Lab Barcelona (Spain), first lab in the European Union, is launched

2007

2006

Neil Gershenfeld presents Fab Labs with a talk at TED[3]

Third Fab Lab meeting in Pretoria (South Africa): The Third International Fab Lab Forum and Symposium on Digital Fabrication [4]

The Center for Bits and Atoms (CBA) is launched at MIT by Neil Gershenfeld: the center investigates the intersection of information to its physical representation, or how bits and atoms interact

2001

1998

Neil Gershenfeld starts to lecture the "How To Make (Almost) Anything" course at MIT, from which the Fab Lab concept will emerge

The first American Fab Lab outside MIT is launched at South End Technology Center in Boston

2003

2002

The first Fab Lab outside MIT is launched at Vigyan Ashram in Pabal (India)

Book:
Walter-Herrmann, J.,
2013. FabLab:
of machines, makers
and inventors,
Bielefeld: transcript.

Ninth Fab Lab meeting
in Yokohama (Japan):
FAB9 The Ninth
International Fab Lab
Forum and Symposium
on Digital Fabrication [12]

2013

Seventh Fab Lab
meeting in Lima
(Peru): FAB7: The
Seventh International
Fab Lab Forum and
Symposium on Digital
Fabrication [10]

Wiki of Fab Lab Iceland
is launched: for a
while this represented
a shared place for
documentation [11]

2011

2010

Sixth Fab Lab meeting
in Amsterdam
 (The Netherlands):
FAB6: The Sixth
International Fab Lab
Forum and Symposium
on Digital Fabrication [8]

First low-cost
Fab Lab launched in
Amersfoort
(The Netherlands)

Launch of the
International Fab Lab
Association [9]

2012

Eighth Fab Lab meet-
ing in Wellington (New
Zealand): FAB8 The
Eighth International
Fab Lab Forum and
Symposium on Digital
Fabrication

Book: Eychenne, F.,
2012. Fab Lab: L'avant-
garde de la nouvelle
révolution industrielle,
Limoges: FYP Editions.

Eleventh Fab Lab
meeting in Boston
(USA): FAB11 The
Eleventh International
Fab Lab Forum and
Symposium on Digital
Fabrication [15]

2015

2014

The fablabs.io platform
for mapping and
coordinating the
Fab Lab network is
launched [13]

Tenth Fab Lab meeting
in Barcelona (Spain):
FAB10 The Tenth
International Fab Lab
Forum and Symposium
on Digital Fabrication [14]

Thirteenth Fab Lab
meeting in Santiago
(Chile): FAB13
The Thirteenth
International Fab Lab
Forum and Symposium
on Digital Fabrication:
Fabricating Society [17]

2017

2016

Twelfth Fab Lab
meeting in Shenzhen
(China): FAB12
The Twelfth
International Fab Lab
Forum and Symposium
on Digital Fabrication [16]

2018

Fourteenth Fab Lab
meeting in Toulouse,
Paris, and everywhere
in France: FAB14
The Fourteenth
International Fab Lab
Forum and Symposium
on Digital Fabrication

Websites:

[1] http://cba.mit.edu/events/05.01.fab/index.html
[2] http://cba.mit.edu/events/05.07.Norway/index.html
[3] http://www.ted.com/talks/neil_gershenfeld_on_fab_labs
[4] http://cba.mit.edu/events/06.06.ZA/symposium.html
[5] http://cba.mit.edu/events/07.08.fab/
[6/11] http://wiki.fablab.is/
[7] http://cba.mit.edu/events/09.08.FAB5/
[8] http://cba.mit.edu/events/10.08.FAB6/index.html
[9] http://www.fablabinternational.org
[10] http://cba.mit.edu/events/11.08.FAB7/index.html
[12] http://www.fab9jp.com
[13] https://www.fablabs.io
[14] https://www.fab10.org
[15] https://fab11.fabevent.org
[16] https://fab12.fabevent.org
[17] https://fab13.fabevent.org

Fab Labs: A beginning

1• Anon, 2012. 'The Third Industrial Revolution'. The Economist. Available at: http://www.economist.com/printedition/2012-04-21

2• Anderson, C., 2012. Makers: The New Industrial Revolution, Crown Business.

3• Gershenfeld, N., 2005. FAB: The Coming Revolution on Your Desktop: From Personal Computers to Personal Fabrication, Basic Books.

It took only twenty years for most human activities to migrate into the digital world. Now that social media and online exchanges have come to be considered fundamental, digital manufacturing technologies are redefining modes of conception and production in our societies. The invention of screen-operated tools and manufacturing processes has, at the dawn of the 21st century, given way to what some describe as a Third Industrial Revolution. According to the Economist newspaper (in its April 2012 special report 'A Third Industrial Revolution')[1], the manufacturing world already has a new face. The old model of mass production has been submerged by the digital wave, and the logistics of very small scale production are slowly defining a new productive order, one that is flexible, distributed and decentralized. Many authors have declared that the world has entered a transitional phase, proclaiming the end of mass production and the arrival of the 'new industrial revolution'[2]. Such predictions are rooted in the current development of new machines for digital self-manufacturing, many of which are used in Fab Labs ('fabrication laboratories'[3] accessible to the public) and other makerspaces.

The 'democratization of production' described by the Maker and Fab Labs movements' spokespersons is accompanied by a 'democratization of innovation'. In a Fab Lab, anybody can invent, create or modify '(almost) anything', with few constraints. Fab Labs, in their approach and the way they defend their practices, are at the forefront of promoting a new mode of individualized designing and making, one which goes against mass cultural tendencies.

Magic lamps, mysterious stones, and other systems capable of turning our dreams into physical realities have fueled a great many science fiction stories. In the 1980s, the replicator taken onboard the spaceship in *Star Trek* perfectly embodied the idea of a machine capable of creating anything you ask it for. Contemporary digital culture enables digital data to actually 'change into' physical objects. The idea of an 'almost magical' machine capable of creating anything is an important aspect of the discussion around the question of dissemination in the world of Fab Labs.

This chapter covers the still fresh but already extremely rich history of Fab Labs. From its incubation at the Massachusetts Institute of Technology (MIT) in Boston, to its spread to every corner of the planet, the Fab Lab movement has displayed unpredicted and multifaceted growth. This global development has involved many people, bringing their local struggles into a movement that has come to receive widespread media attention.

While researching for a thesis on this very subject, I traveled the world visiting digital manufacturing collective spaces. I visited two pioneer Fab Labs, one in Norway, way above the Arctic Polar Circle, and the other in the heart of Boston. This is the story of those two visits, alongside selections of interviews and drawings from my field notebooks.

A food duplicator as seen in the original series of *Star Trek*, here reproduced by 3D artist Shir Shpatz.

How to make (almost) anything: the first steps of digital self-manufacturing

The history of Fab Labs and of personal digital fabrication as we know it today started at the Center for Bits and Atoms at MIT, in the late 1990s and early 2000s.

A 'useful' course

In 1998, professor Neil Gershenfeld offered MIT students a one-semester course called 'How To Make (Almost) Anything'. The laboratories of the Center for Bits and Atoms were generously equipped with the latest generation of lasers, water jet cutters and microcontrollers. In order to expand his research into digital self-manufacturing, Gershenfeld decided to open a lab to a handful of students for a practical initiation into how to use these machines.

The original idea was simple: to use very complex machines to make other machines. Instead of obliging students to spend a year trying to understand how to

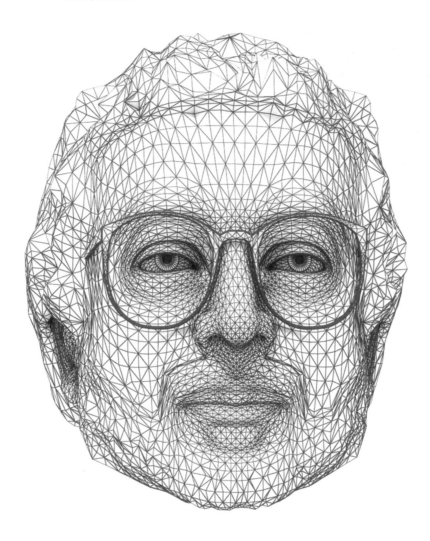

Neil Gernshenfeld in a 3D
portrait by Igor Lobanov.

approach these complicated systems, he decided to establish a series of guidelines for
a small group of students at the end of their course, to help them finalize their research:
'Imagine our surprise, then, when a hundred or so students showed up for a class that
could only hold ten [...] They weren't the ones we expected, either; there were as many
artists and architects as engineers. And student after student said something along the
lines of "All my life I've been waiting to take a class like this," or "I'll do anything to get into
this class." Then they'd quietly ask, "This seems to be too useful for a place like MIT – are
you really allowed to teach it here?"'

4• Ibid.

    In FAB, The Coming Revolution on Your Desktop[4], Neil Gershenfeld details the be-
ginnings of his research, which became a calling into question of classic teaching methods inside one of the best universities in America. 'Virtually, no one was doing this for
research. Instead, they were motivated by the desire to make things they'd always wanted,
but that didn't exist,' Gershenfeld explains.

The students who took the course in the first year were rather good at art and creation and didn't know much about engineering. Still, they all succeeded in finalizing an original and functional system, from their 'intelligent' objects' external form – which required the use of computer-operated machines – to internal functions – which called for the assembly of the associated electronic circuit. The course (still taught at MIT) is based on the demand for, rather than the supply of, knowledge, since the students' needs are addressed according to the requirements of each project.

Many crazy objects were made during that first year, some of which remained famous and are often put forward to illustrate the first steps of digital personal fabrication. There is, for example, an alarm clock that needs to be punched to prove one is awake, or a web browser made to allow parrots to communicate with each other around the world. The feature common to these products is that they are destined for a one-person market; they were not commissioned, and fulfill a personal, non-professional desire. Such is the case of the project executed by Kelly Dobson, a young artist with little electronic knowledge but whose creation has become emblematic of that first experience: it is a bag that you can put over your mouth, allowing you to scream in public without disturbing the people around you. Kelly felt that in many situations the presence of other people kept her from expressing certain emotions, such as screaming out of rage or exhaustion. ScreamBody was designed to trap all sound when she screamed into it. Later, by applying pressure on the bag, she 'frees' it of its contained, recorded noise.

A tool for personal expression

The video that presents the object shows her in a crowded subway train, silently screaming into her bag. The following sequence shows her on the street, pushing her scream out of the bag, which recorded it by way of a system of tiny integrated speakers. In her video, Kelly goes over every aspect of the object's construction: the bag itself, the electronic circuit, the associated program, the recording system and the sensors. 'Kelly sees the design of circuits as an aspect of personal expression, not product development,' explains Neil Gershenfeld; '[S]he didn't design ScreamBody to fill a market need; she did it because she wanted one.' The 'killer app' of digital self-manufacturing – as it was prefigured in the early 2000s, during the first editions of Gershenfeld's famous course – was about creating products for a one-person market.

Stepping outside the walls of MIT

'All of this happened by accident'

This university adventure features in the history of the movement as a foundation stone. The global dissemination of that educational experiment, and the emancipation and disinhibition of technological logistics that it advocated, started a movement that now includes over 1000 Fab Labs around the world.

Exploring 'the implications and applications of personal fabrication in those parts of the planet that don't get to go to MIT': this was according to Neil Gershenfeld the motivation behind the transition from his course's educational experimentation to the implementing of the first Fab Labs outside the university. But the MIT team hadn't anticipated such a success. 'We did not have an agenda for such a development. We did not plan all that and the growth of the network after the first Fab Labs is far above our wildest expectations. It was an accident!'

The tipping point of the Fab Lab adventure was to a large degree the moment that the National Science Foundation (NSF) granted the Center for Bits and Atoms financial support for their research. The counterpart of that grant was an appreciation of the Center's progress in more ordinary fields, equipping effectively different world populations with machines tested in the prestigious university. Starting in 2002, a first wave brought Fab Labs to India, Costa Rica, northern Norway, Boston and Ghana, with an average budget of $20,000 for each lab. These first labs were not, then, destined to be financially independent, but were supported by MIT. Teams composed of students and researchers were sent into the field. These teams were motivated by the fact that the 'digital fracture' between richer and poorer countries is not only a question of access to computers, but a more long-term issue. Sending computers all across the world is thus not a solution: it is more worthwhile to send the components needed to make computers, in accordance with the realities and needs of each location. The young researchers around Neil Gershenfeld at MIT adhered to a progressive discourse: 'Instead of building better bombs, emerging technologies can help building better communities.'

The development of the first Fab Labs relied on certain community leaders, public figures already involved in the development and animation of local communities. The 'combination of need and opportunity, [led] these people to become technological protagonists rather than just spectators'; among them are Mel King in Boston, Haakon Karlsen in Lyngen (Norway) and Kalbag in Pabal (India), three founding figures whose names might have been forgotten due to recent relocations and the exponential development of the whole network. The history of Fab Labs is often revised and some essential aspects of it are sometimes forgotten, among them these humble, very local actions, which kept up their momentum in the face of technological challenges.

The Fab Lab movement rapidly extended beyond MIT, as certain spaces sought advice on how to set up the same type of digital manufacturing workshops. The need to establish a global charter stemmed from this expansion. The Fab Charter[5] – which has grown since its birth and has been translated into several languages by Fab Lab members – has to be posted up in each space that considers itself part of the Fab Lab group. It recalls some essential definitions: 'Fab Labs are a global network of local labs, enabling invention by providing access to tools for digital fabrication.'

Fab Labs share an evolving inventory of core capabilities to make (almost) anything, allowing projects to be shared. The Fab Lab network is described as being able to provide '[O]perational, educational, technical, financial, and logistical assistance beyond what's available within one lab'. According to the charter, 'Fab Labs are available as a community resource, offering open access for individuals as well as scheduled access for programs'. Certain philosophies and responsibilities are also detailed in the charter's text: 'Safety: not hurting people or machines. Operations: assisting with cleaning, maintaining, and improving the lab. Knowledge: contributing to documentation and instruction.'

I was able to interview Sherry Lassiter – director of the Fab Foundation[6] – about the foundations of the charter. She has been committed to the deployment and development of Fab Labs around the world since the very first years. 'We like the fact that all Fab Labs are different. It has to come from the community. You can't come and say, "Here is your Fab Lab". Fab Labs have more impact when they are built by communities themselves,' she told me.

The charter's take on intellectual property has given rise to a few debates and different interpretations: 'Designs and processes developed in Fab Labs can be protected and sold however an inventor chooses, but should remain available for individuals to use and learn from.' We also learn that 'Commercial activities can be prototyped and incubated in a Fab Lab, but they must not conflict with other uses, they should grow beyond rather than within the lab, and they are expected to benefit the inventors, labs, and networks that contribute to their success.' Sherry Lassiter admits the text is vague, but explains that it was 'very controversial when we first started, when we said everything in the lab had to be open source. Lots of startups or entrepreneurs got scared. So we drove back away from that and said "OK, deal as you want with intellectual property, but we are public spaces so you have to give back to us in some way. If we help you make money and make a business, help us back. We want some reciprocity in there, but we can't dictate." And I think that the network is mostly copyleft rather than copyright, now.'

The Fab Charter

5• CBA, 2012. The Fab Charter. Center for Bits and Atoms. Available at: http://fab.cba.mit.edu/about/charter/

6• Editor's note: the Fab Foundation was established in 2009 to facilitate and support the growth of the international Fab Lab network. The Foundation works mainly at global level and as a 'wrapper' among the many regional and national Fab Lab organizations. Anon, Fab Foundation. Available at: http://www.fabfoundation.org/

Vigyan Ashram: a farm in India

It is at Vigyan Ashram, a few kilometers away from Mumbai, India, that the first Fab Lab in the world was set up in the early 2000s. The place had been a rural school in Western India, in the middle of a small village called Pabal. There, under the leadership of Kalbag – previously a professor of nutrition technologies – groups of young people receive scientific training directly applied to the living conditions of their own region, a poor and dry environment with very limited access to water. Based on the principles of teaching through practicing, the school was built with these very students, children who had dropped out of the classic school circuit. Vigyan Ashram is a fee-paying school that slowly became independent thanks to funding from small enterprises developed around the school to measure and localize water points. The construction of tractors using pieces of dead Jeeps (known as MechBull) also brings in a little money. Before the arrival of the MIT teams, Pabal had already been a surprisingly innovative town for some years, thanks to Kalbag and the small community he had built around his projects. A single visit from the group formed around Neil Gershenfeld was enough to connect activities already set up by the local school team, on a very limited budget, with tools developed simultaneously by researchers at MIT: 'When I first met Kalbag, I casually mentioned the kinds of personal fabrication tools we were developing and using at MIT […] He reeled off a long list of things they wished they could measure, but weren't able to,' Gershenfeld recalls. Kalbag and his students have survival on their minds more than profit, and their main concerns are related to agricultural issues. In a dry, sparse rural context, the need to obtain certain data from the environment is crucial. That way, the instrumentation required for the management of water, milk, rice, eggs and other essential products guarantees a form of energy independence, as well as long-term development.

'Because Kalbag and his students had to produce both the food and technology that they consumed, the impact of access to tools to develop instruments was much more immediate for them than for an engineer surrounded by support systems,' Gershenfeld explains; '[T]he recurring rural demand for analytical instrumentation led to an easy decision to launch a Fab Lab with Kalbag at Vigyan Ashram, to be able to locally develop and produce solutions to these local problems. The opportunity in Pabal was clear; less expected was the impact of the lab outside of the region.' As a result of a discussion between the MIT team and Kalbag's team about Indian electricity networks and energy losses due to the theft and hijacking of electricity in the region, the very first Fab Lab was born. A short time after that encounter, it became an experimentation space for engineers to come from Delhi to develop prototypes of measuring tools for the local electricity network. A few months later, back in the United States, the situation reversed itself: Neil Gershenfeld was visiting the engineering development lab of an important society working on the research made by MIT. Faced with the great inertia of this society's infrastructure regarding tests and product trials – which pass through the hands of many engineering teams – Gershenfeld proposed bending the protocols by testing these ideas back in Kalbag's Fab Lab in Pabal, in faraway India. 'When I jokingly suggested that we go to Kalbag's farm, the silence in the room suggested that the joke was very serious.'

'Growing Inventors' is the chapter in FAB[7] in which Neil Gershenfeld presents those community pillars on which Fab Labs were supported; exceptional people committed to local causes in the service of isolated and marginalized populations. After Pabal's Fab Lab, Boston's Fab Lab in the South End Technology Center opened its doors in 2003, with the help of another well-known figure already active in his own community, for very different reasons: Mel King. In 2004 it was Ghana's turn, with the Takoradi Technical Institute, which in its first years attracted many children from the streets, and then divided its activities between the teaching of some software and the development of 'real-world applications', mostly for solar energy.

7• Gershenfeld, 2005.

Dr. Shrinath Kalbag, founder of Vigyan Ashram, India. Portrait by Fabio La Fauci

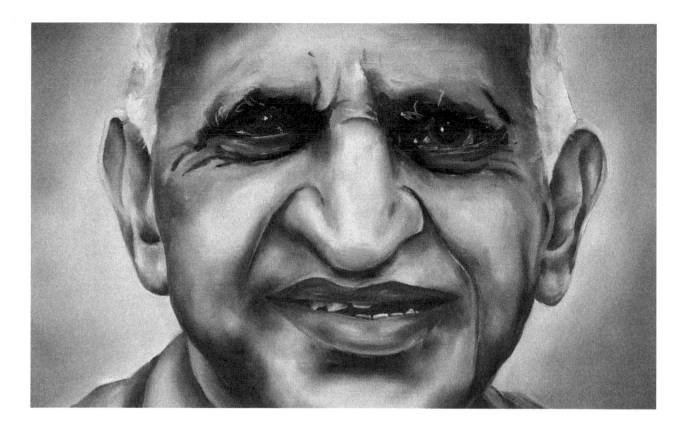

All the way from Boston to Lyngen

South End Technology Center, Boston

The Fab Labs at South End Technology Center and Vigyan Ashram are two examples that reveal a whole side of the Fab Lab movement's history, dating from before its official birth. The primary ambitions of these pioneering spaces are well characterized. They range from solving essential environmental issues to community gathering and digital training. In the first version of the project, the Fab Lab network was to be made up of workshops scattered around the world, each one equipped with the exact same machines. That principle offered each Fab Lab, wherever it was, the possibility of replicating identically or adapting locally any project produced elsewhere in the network. This almost standardized design for the spaces was not to last. It wasn't long before different labs started making different choices, with each Fab Lab looking before anything else to offer local communities the access to digital manufacturing tools.

In April 2014 I traveled to the South End Technology Center. This is the first American Fab Lab, set up in Boston in 2003. The goal of my visit there, over ten years after its birth, was to meet Mel King, a man who marked Boston's history and prefigured this Fab Lab's. This peculiar space combines his battle to access knowledge and technology with the birth of a movement now in the midst of a full expansion. So I met with Mel King outside the Fab Lab, in the heart of a rather quiet, residential Boston neighborhood. Nothing more than a modest sign indicates that there is a Fab Lab on this street. Mel King was sitting on a bench, in front of the entrance to the South End Technology Center, located in the basement of a brownstone building. Through the windows, I spotted a few young people gathered around a 3D printer. Pamela King, Mel's daughter, caught up with us. She is also involved in Fab Lab development and works at Fab Foundation with Sherry Lassiter. Mel King lives in this very neighborhood; he has limited mobility, but insists on coming to the Fab Lab every day (he was already quite old when he came across the Fab Lab movement). He is renowned for having led the Tent City movement in 1968 in a poor area of the city, to protest against the planned installation of a gigantic parking lot instead of housing. Convinced that city inhabitants could influence the decision by protesting, he contributed to gathering people to occupy the land and installing hundreds of tents in the zone. This episode led to the construction of a social housing project called Tent City, in homage to the demonstration that inspired them. Soon after that, Mel King became the founder of the South End Technology Center, a 'community center' that opened in the same area. For about thirty years now, this space has offered training in digital technology.

After he retired in the late 1970s, the former MIT professor invested all his time in arranging and leading the space, which started out solely with computers. King is now over 90 years old, but he still wears overalls and a USB stick hanging from a cord around his neck.

When I asked him about Fab Labs and the international popularity the movement has enjoyed, he answered by reversing the question: 'Technology is not a solution for everything. I believe in technology of the heart. What matters most is what people can share and give to each other, learn from each other.' Mel King made a radical choice when he retired in the 1990s. Senior researchers at MIT traditionally have a right to keep an office on campus to pursue their research. Instead, King asked if he could transpose his office into the heart of Tent City and share it with the neighborhood community.

'At MIT, we had very high-quality equipment. When I saw Internet coming, I was worried that people with low income might not have any access to this technology. I did not want that to happen.' That's how the space was born, focusing as it did mainly on computers. 'When I retired it was the time when people could go to Internet cafés, in 1996. I settled here the first open space dedicated to computers in Boston. MIT pays the rent. In the early days, we were doing a thing called "midnight computer", when we were opening very late. We were working with a girl who studied for her PhD at Media Lab with Seymour Papert. We would go in the street, show people how to fix their computers, how to build things. Then I met Neil Gershenfeld and we realized that we were doing similar things, him with his Fab Labs and myself with the local people here.' Gradually, the South End Technology Center was equipped with digital machines and eventually came to be called a Fab Lab in 2003, after years of fighting for community access to knowledge and technology.

'In the beginning we did not have all the equipment that interested Neil, but there was something in his program that was very compatible with what we had been doing from scratch for years. In fact, nothing is new in this Fab Lab's story. Teaching children how to make things? We have been doing that for years, except now we can use a 3D printer instead of wooden pieces.' Mel King answered my questions regarding the Fab Lab network and history very vaguely and allusively, making it a point to focus on the more fundamental values that pushed him to keep the space open to everyone, for free. A noticeable difference: instead of the usual Fab Charter, the UN Charter of Human Rights is posted by the entrance.

It is as if the arrival of digital manufacturing machines a decade ago didn't much change the DNA of the space. The 'Learn to teach, teach to learn' program has been in place for several years and gathers young people from the area around projects that go from Excel spreadsheet management to poetry practice, or from 3D printing to digital embroidery (the latter is often used by neighborhood teenagers to customize their hoodies). 'In the USA,' Mel King explains, 'technology is often used to produce weapons. Army is not progress. What kills men is not progress, it's not "high tech". It's more "low" tech or no tech at all. Technology should encourage the power of life.'

MIT-FabLab Norway, Lyngen

The MIT-FabLab Norway is one of few in the world that include the MIT acronym in its name. Its birth in 2003 – shortly after the Boston Fab Lab – mixed the evolution of a farm on the Lyngen fjord, the commitment of Boston's greatest engineering researchers, and the development of local sheep artificial insemination techniques. I traveled to Lyngen in June 2013 and spent ten days on site, where I was able to talk at leisure with Haakon Karlsen Jr., who – alongside Mel King and Kalbag – was one of the pillars of the movement's first years.

Getting to MIT-FabLab Norway requires a long journey: from Olso to Tromsø and finally Lyngen. Housed in a vast cabin, the entrance to the land it stands in is marked by two flags: one from the region, and one from the USA. The main cabin is surrounded by smaller ones for accommodation. About 600 people pass this Fab Lab's door every year. When I arrived at around 9 pm, it was raining lightly. I left my hiking backpack in the entrance, took my raincoat and shoes off and met Haakon Karlsen Jr., his wife Gunn, Junior the dog, and two women in the kitchen who were just taking out of the oven a huge dish of fish fresh caught in the fjord that very morning. We sat right down to dinner. Haakon sat on one of the chairs designed especially by Jens Dyvik – a young globetrotting designer also known as the 'Fab Lab nomad', who recently traveled the world visiting each and every Fab Lab, leaving a few designs behind him[8].

8• Editor's note: the history of Jens Dyvik's Layer Chair can be read in Chapter 6.

Haakon is in his early sixties. He was born in Lyngen and, after training in engineering, spent some of his youth working on sheep insemination in the family farm right below the land where the Fab Lab was later installed. A key figure in the area, he was also a teacher and then a farmer. He owns several houses and pieces of land along the fjord. Roughly a decade ago, he contributed to shaping the outline of the Fab Lab network with MIT. He sports a gray goatee and Crocs, and has a round belly. He loves telling stories, in particular his own version of the Fab Lab story and that of his own Fab Lab. 'It all started a little before the year 2000,' he recalls, '[T]here were many diseases and it was necessary to boost growth in some herds. In 1994, the Norwegian government was asked to establish a laboratory for artificial insemination of sheep, deer and goats. With some farmers and shepherds in the region, we got surprising success rates of up to 94 % instead of the usual 10 %. We quickly realized this was due to two farmers we were working with who knew their animals well and knew how to inseminate at the exact moment of ovulation. To succeed, it was necessary to know when the females were in heat. I suggested that we imagine for ourselves a technical tool to measure hormones.' So the premises of the Norwegian Fab Lab emerged from the coming together of a pragmatic need in the region and Haakon Karlsen's engineering skills. The great vault in the Fab Lab gives the impression of a strange chapel in which one must talk very softly. In the conversations I recorded with Karlsen, silences play an important role. We can hear the wind blowing outside and imagine the soft light of the unsleeping summer sun.

'We tried to detect different hormones to see what could be learned. We finally developed a small machine that captured the temperature and sent a message to warn the farmer about the time of ovulation. It was based on the female brain activity curves. Then we created a program to educate shepherds about the tool. Later, with farmers, we thought about a possible use for the rest of the year. So we put an accelerometer in our little machine to capture the movements of the sheep. To test this feature, we created a system that calls home after fifteen minutes of inactivity for the sheep, saying, "I'm dead". We then put in a GPS, which allowed us to get the geographical coordinates of the sheep sent to the farmers.' The Electronic Shepherd project allows a shepherd to localize sheep flocks in the mountains in order to protect them from wolves and unstable terrain. Back when that research was carried out, the Fab Lab as we know it today didn't yet exist. Still, the 'lab' installed in the farm was already equipped with electronics.

'It had everything to do with welding [...] That's where we got the idea for the sheep phone. But it was difficult to get the signal from the mountains to the farms. We worked with Telenor [a Norwegian telecommunications company] for one year.' With this project, and thanks to the help of the National Science Foundation, MIT ended up noticing Haakon Karlsen's team. 'There was an innovation competition launched by MIT globally to develop local projects. MIT sent some of its best teachers to Norway to find a suitable cooperation project. They found us through Telenor, who told them: "There is this crazy guy lost in the fjord who devised sensors for his animals [...]" We enjoyed a great year of cooperation with MIT in 2001 and we were invited to Boston to present and develop this project.' The team then consisted of Haakon and his son Jurgen, who worked on the farm. 'It was fantastic,

but after years of collaboration we had to terminate the project. We had a discussion at MIT in Boston and we decided to do something to further enable this kind of adventure, something we would call [...] a Fab Lab. A Fabrication Laboratory. The decision was taken on 18 October 2002, I remember. We first decided to launch three Fab Labs. One in Pune with a man named Kalbag, from Vigyan Ashram, south of Mumbai, and another in a poor neighborhood of Boston called South End Technology Center, with Mel King. And the third here in Norway. At first we did not really know what we were doing. The definition at MIT was "rapid prototyping". But since then, things have changed and other places are born with other definitions. In my opinion many Fab Labs now exist that just have the name Fab Lab [...] My definition? "A global network of people who want to work together and share their knowledge." That's all.' Haakon Karlsen tells the story his way. He believes that today, his Fab Lab is more a 'community center' than it is a prototyping space: 'We even celebrated a wedding here!'

The arrangement of the machines, tables and workstations in the main room of the cabin makes this clear immediately. The whole technical aspect is now on the outskirts, along the walls, set aside. In the center sits a large table for reunions and videoconferences, and then a huge chimney, several tables for meals and a few armchairs. The open kitchen in itself takes up a lot of room. Karlsen jokes about it readily: 'When Neil Gershenfeld of MIT came to see the finished chalet and saw the kitchen, he told me that it was useless, that I had made a mistake, that it was not planned! The result proved that I was right. A Fab Lab is people, not just machines.' Coffee, a choice of tea, muesli, cookies and local firewater are made available. Some of the tables are already set to welcome potential visitors, who often wish to stay in the area for a few days for hiking or other outdoor activities. These days, the Fab Lab is both an inn of sorts and a prototyping and manufacturing space, the latter aspect accounting for a significant portion of its income.

'In 2004 we built this house. All the equipment came from Boston, free. Why here? Good question, ultimately. We must ask Gershenfeld or Sherry Lassiter. Initially, the Lab was down on the farm. I am not an architect, but I made all the plans. When the house was built, we installed all the machines here. Then MIT sent other machines and some students. Neil came, his wife, his twins, as well as Sherry Lassiter and Amy Sun. Engineers, researchers, who were there to install the machines with my son and me. It was great. Then, they traveled everywhere to settle other places like that. But I know that in their heart, the Norwegian Fab Lab is really special.' Symptomically of the friction that exists between the rural and technological worlds, the big digital milling machine is not in the Fab Lab's main cabin: it has been installed at the farm, hidden behind a door in the back of a cluttered barn, a kayak hanging from the ceiling. In the winter, when the sheep are back inside, they co-exist with the milling machine in a happy mess.

Nowadays, the daily activity of this pioneer Fab Lab is quite limited, aside from when special workshops use the space, bringing together key figures of the network and MIT students. The rest of the time it looks more like it did when I visited: the machines are off, Haaken Karlsen is either at home or in the Fab Lab in front of his computer, and people pass by to have coffee, inquire about the flocks or the impending birth of a foal in the nearby field, or to repair something.

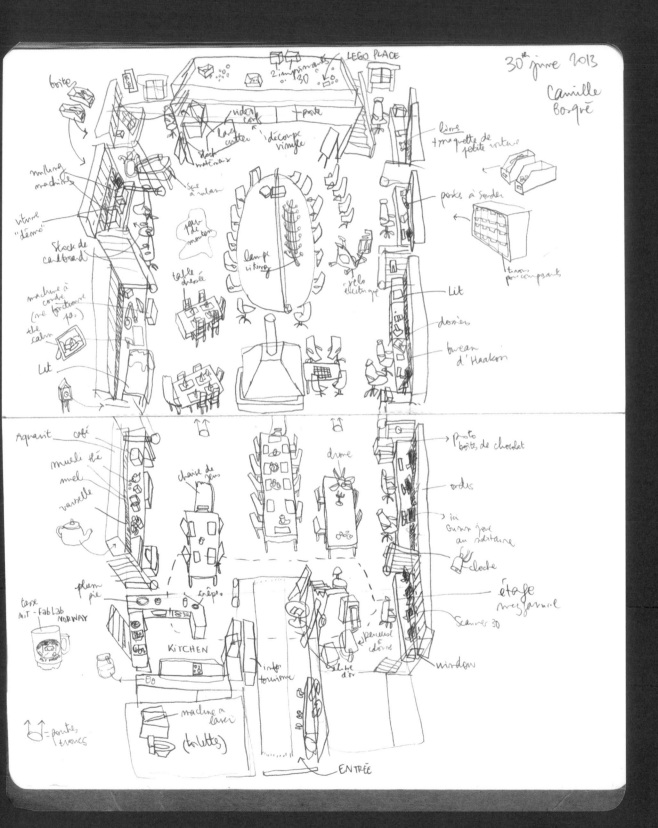

LEGO PLACE

30th juin 2013

Camille
Bosqué

$\uparrow\uparrow$ = portes franches

Opposite page:
FabLab Norway, Lyngen,
Camille Bosqué.

All is FAB

Of the more than 1000 Fab Labs on the official list at the time of writing, many spaces are still growing and may not be as active or internationally connected as others. Language barriers, isolation, or problems with Internet connections can sometimes constitute obstacles that keep the dream of a hyper-connected network from being fully realized. Installed in rented or squatted premises, in basements, school facilities, museums or libraries, inserted in professional training spaces, built as associations, dependent on organizations or corporations, the modes of existence of Fab Labs are now extremely diverse, responding as they do to diverse local situations.

French Fab Labs are a particular case among this network. At present, France boasts close to 140 Fab Labs on the movement's official website, and is named as the country with the second largest number of Fab Labs after the United States. But 'French Fab Labs are an exception that we do not understand, they are really apart,' says Sherry Lassiter. France is indeed something of an outsider. It redefines the very concept of Fab Labs, and doesn't hesitate to associate the name with any digital manufacturing workshop, whether or not it is open to the public or included in the global network.

Official variations run wild and take the concept of Fab Labs further, from Fab Foundation to Fab Connection[9], Fab Academy[10] or even Fab City[11]. I asked Sherry Lassiter if she thought the name Fab Lab should become a brand. Her answer was decisive: 'I think in a way Fab Lab is already a brand, it's an identity. We try not to copyright it. In the Netherlands, they did, they licensed it for €1 a year and it gives them the right to say [...] you cannot have the license. Spain just did the same thing and trademarked the logo. I think it's OK. They were afraid that commercial enterprises might take all that we build and make it their own, taking away our reputation. We have hesitated a lot. It's a little scary sometimes, to think that people are going to abuse our name and network identity. But we don't want to make a brand.' What is the future of the movement? According to Neil Gershenfeld, whom I also interviewed on the subject, '[I]t's hard to say, because we only set up one lab from MIT and we are approaching 400 labs now' (this total has risen to more than 1000 since this interview was made). Since its creation, the network applies 'its own Moore's law and doubles its size every year'. In the future, the phenomenon should be able to be amplified, the dream of MIT's team being that 'each Fab Lab could create other Fab Labs'. The real project is organizational, Gershenfeld asserts: 'Here in Barcelona, the city is inventing a whole new urbanism, where each citizen should have access to tools that will help Barcelona become self sufficient.'

9• Editor's note: Fab Connection creates business opportunities for individuals and labs in the Fab Lab network by offering a job marketplace, team building services and promotion of educational activities. Anon, FabConnections | Facilitating growth in the FabLab Network. Available at: http://www.fabconnections.org/

10• Editor's note: the Fab Academy is a Digital Fabrication Program distributed among many Fab Labs and directed by Neil Gershenfeld. It is based on MIT's How to Make (Almost) Anything course, now offered outside MIT in the Fab Lab network. Anon, Fab Academy. Available at: http://www.fabacademy.org/

11• Editor's note: the IAAC | Fab Lab Barcelona is developing the concept of a Fab City, a smart city not only managed by technology and companies, but also by projects developed by Makers and Fab Labs. You can read more at: http://fab.city.

Massimo Menichinelli

Anatomy of a Fab Lab

Technologies, tools and materials available in Fab Labs

Fab Labs are communities, services, places, and - especially - technologies: their initial goal was the democratization of access to new digital fabrication technologies and processes. Over the years, a specific set of technologies has come to make up the inventory for Fab Labs. Even if individual Fab Labs are free to make specific choices regarding brand, power and features, technological typologies are important in enabling each lab to reproduce projects developed in other labs. New machines and processes, the majority of them digital, are increasingly becoming part of Fab Lab practice; all of the possible technologies cannot, however, be found in each Fab Lab. This chapter covers a selection of the most widely available technologies - and a few emerging ones - that can be found in Fab Labs.

Technologies, tools and materials

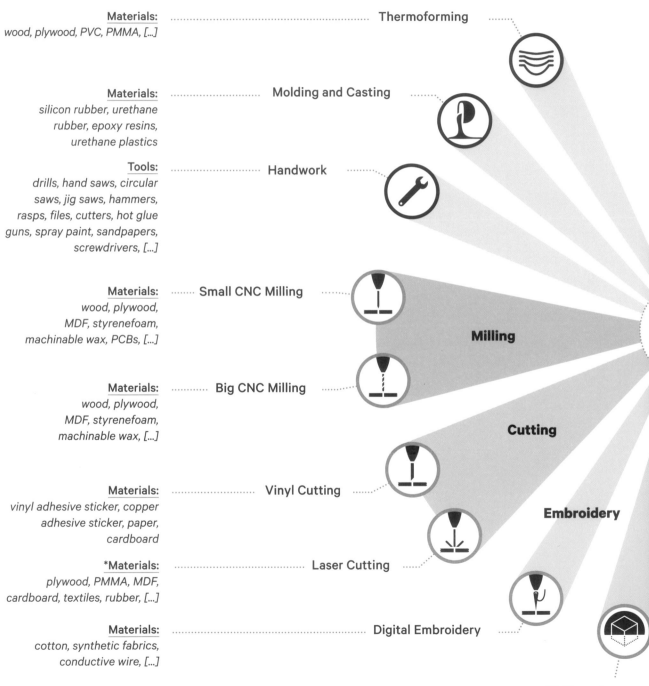

Materials:
wood, plywood, PVC, PMMA, [...]

Thermoforming

Materials:
silicon rubber, urethane rubber, epoxy resins, urethane plastics

Molding and Casting

Tools:
drills, hand saws, circular saws, jig saws, hammers, rasps, files, cutters, hot glue guns, spray paint, sandpapers, screwdrivers, [...]

Handwork

Materials:
wood, plywood, MDF, styrenefoam, machinable wax, PCBs, [...]

Small CNC Milling

Materials:
wood, plywood, MDF, styrenefoam, machinable wax, [...]

Big CNC Milling

Milling

Cutting

Materials:
vinyl adhesive sticker, copper adhesive sticker, paper, cardboard

Vinyl Cutting

Embroidery

***Materials:**
plywood, PMMA, MDF, cardboard, textiles, rubber, [...]

Laser Cutting

Materials:
cotton, synthetic fabrics, conductive wire, [...]

Digital Embroidery

3D Scanning

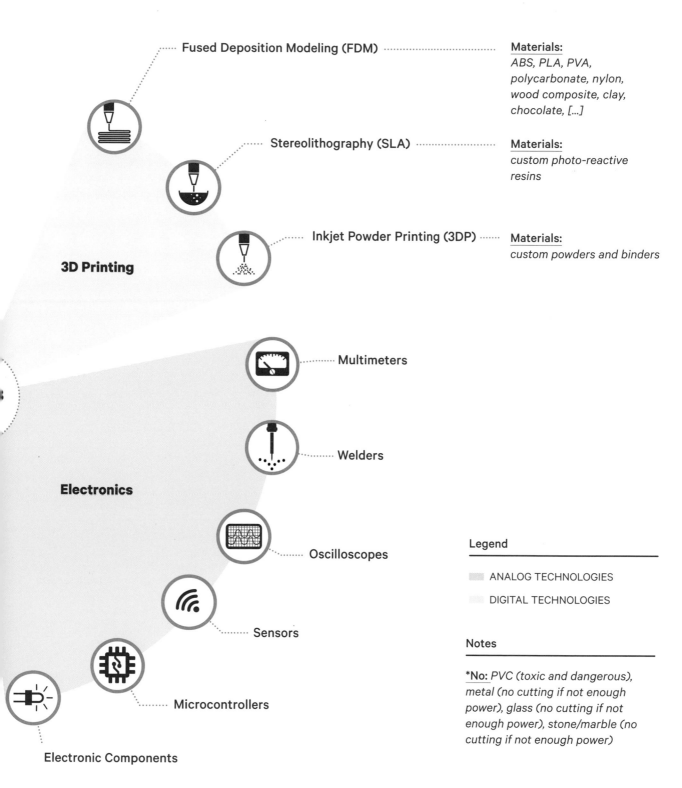

3D Printing

Fused Deposition Modeling (FDM)

Materials:
ABS, PLA, PVA, polycarbonate, nylon, wood composite, clay, chocolate, [...]

Stereolithography (SLA)

Materials:
custom photo-reactive resins

Inkjet Powder Printing (3DP)

Materials:
custom powders and binders

Electronics

Multimeters

Welders

Oscilloscopes

Sensors

Microcontrollers

Electronic Components

Legend

ANALOG TECHNOLOGIES

DIGITAL TECHNOLOGIES

Notes

*No: *PVC (toxic and dangerous), metal (no cutting if not enough power), glass (no cutting if not enough power), stone/marble (no cutting if not enough power)*

Electronics production

TECHNOLOGY

The electronics production facilities in a Fab Lab consist of a set of different machines, tools and components. Instead of a single machine creating an object, it is the integration of different tools and components that create the electronics.

Oscilloscopes are used to observe the changes of an electrical signal by plotting voltage over time on a small screen. With an oscilloscope, we can check circuits for faults, because it allows us to see the signals at different points in the circuit: it is therefore possible to understand the voltage communications and their patterns in the circuit. A multimeter is an electronic measuring instrument that measures voltage, current, and resistance in different parts of the circuits, also allowing us to discover breaks between two points in the circuit.

In order to apply such differences to the circuits, we need to solder some specific electronic components to the circuits. Such components affect the flow of electrical energy and its values, and can be of different types, according to their functions. For example, there are resistors (for the reduction of current flow and voltage levels), capacitors (for the storage and release of electrical charge), antennas (transmission or reception of radio waves), sensors (reaction to environmental conditions like temperature, force, light, distance, humidity and air quality by generating an electrical signal), actuators (motors), microcontrollers (intelligence, memory and logic of the circuit), and so on. In order to attach these electronic components to the PCB, we need to solder them with a soldering iron and a solder wire. The soldering iron melts the solder wire and heats up the components and the circuit, joining them together.

As well as all these components, tools, and machines, electronic production facilities can include many different tools, such as lenses or microscopes for the magnification of smaller components, tweezers, craft knives, tapes, glues, heat guns, and so on.

SOFTWARE AND KNOWLEDGE REQUIRED

The design and manufacturing of electronics requires an understanding of some laws of physics related to electrical currents, knowledge of electronic components and their features, and experience in handling all the tools. PCBs can be designed with software such as the proprietary but also freeware CadSoft EAGLE, or the open source KiCad. Both can design the logic of the schematic of a circuit, and the corresponding physical layout of a PCB. The electronic circuits also need to be programmed; familiarity with programming languages like C, C++ or Assembly is therefore required.

Microcontrollers

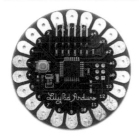

TECHNOLOGY

The electronics production facilities in Fab Labs enable us to build printed circuit boards and electronics projects from scratch, by designing and manufacturing our own small computers. In order to put intelligence inside our PCBs, we need to insert microcontrollers: these are small computers on a single integrated circuit containing a processor core, memory, and programmable input/output peripherals. Sometimes the users of a Fab Lab decide to design their own PCB by designing the connections of the microcontroller with all the electronic components. This option gives the user a wide range of possibilities, but this process takes time, and users in Fab Labs thus tend to prefer pre-built single-board microcontrollers. Single-board microcontrollers consist of a microcontroller built onto a single printed circuit board, providing all the circuitry necessary for the development of a project. Single-board microcontrollers, then, are like small computers that can be extended by the users with electronic components and sensors into full operative devices. Such devices are basically a cheap and easy-to-use physical interface wrapped around a microcontroller, enabling the user to develop her/his own projects without spending time on the minutiae of setting up the microcontroller. There are many single-board microcontrollers available, with different features, prices and degree of openness: a great example is the Arduino boards family.

SOFTWARE AND KNOWLEDGE REQUIRED

Microcontrollers alone cannot do anything: they need to be connected to electronic components and other devices, and they need to be programmed to work with these components and devices. Therefore, besides the knowledge required for electronic production, users need to know one or more programming languages in order to develop the functions of the microcontrollers. Compared to electronic circuits built from scratch, microcontrollers are usually easier to program, and users can use simpler programming languages like simplified C, Python or Processing, besides C, C++, Assembly and so on.

Laser cutting

TECHNOLOGY

Laser cutting is a technology that uses a laser beam to cut or engrave materials; which materials it can cut or engrave depends on the power of the machine and on the characteristics of the materials. Laser cutting machines (or laser cutters) have been used in factories and other industrial contexts for many years, and less powerful (and less dangerous) versions are becoming widespread, especially in Fab Labs. The laser cutters are the most popular machines in most Fab Labs, since they are very versatile and can be used on many materials. Laser cutters are likely to be the most expensive or among the most expensive technologies in a Fab Lab, and for this reason not all Fab Labs start out with a laser cutter. However, due to their popularity, having a laser cutter in a Fab Lab is of critical importance. There are several different technologies for laser cutting, but most of the laser cutters used in Fab Labs are CO_2 lasers. When stimulated by an electric current, CO_2 and the other gases present in the laser tube become agitated, emitting laser light in the infrared and microwave region of the spectrum. Infrared radiation is heat, and this laser basically melts or sets fire to the material it is focused on, through computer numerical controlled optics. The focused laser beam is directed at the material that is melted, burned, or vaporized away. Depending on the speed, power and pulse of the laser, we can achieve different effects on many materials: we can either engrave or cut. Engraving is achieved by transforming the drawing into a bitmap image, made of many 'dots' (pixels): the laser head scans back and forth, engraving a series of dots one line at a time. Engraving then works exactly like an inkjet printer, except that it burns pixels instead of leaving ink drops as pixels. Cutting, instead, is based on the movement of the laser head that always follows a continuous path. This path needs to be designed or converted as a vector drawing, not as a bitmap, as with engraving. Many different materials can be laser cut or engraved, but the possibility of cutting through (and the maximum thickness of the materials) depends on the power of each specific machine. Most of the laser cutters available in Fab Labs can engrave but not cut metals, glass or stones, for example. Usually Fab Lab managers forbid users to laser engrave or cut PVC (or plastics containing PVC) because when lasered and burnt they emit molecules that are toxic for people and machines.

SOFTWARE AND KNOWLEDGE REQUIRED

Besides learning how to operate a laser cutter, this technology only requires knowledge of creating 2D designs: vector drawings for cutting and bitmap (raster) drawings for engraving. These drawings can be created with many types of software; for example, vector drawings can be created with proprietary software Adobe Illustrator or open source software Inkscape; raster drawings can be created with proprietary software Adobe Photoshop or open source software Gimp. Vector drawings can easily be converted to raster drawings; going in the opposite direction is possible but more difficult.

Vinyl cutting

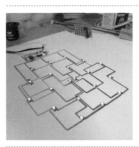

TECHNOLOGY

Since cutting vinyl or PVC with a laser cutter is prohibited for health reasons, this is covered in Fab Labs by vinyl cutters. Such machines have a very simple structure and function: a cutter blade is moved along a continuous path, a vector drawing, cutting the roll or the sheets that are inside the machine. With this technology we can cut paper, cardboard and especially vinyl stickers. After cutting the shapes with the vinyl cutter, we usually need to remove manually the parts that are not needed, and use a transfer sheet for moving the sticker onto the final surface.

Vinyl cutters are mainly used to make signs, banners and advertisements. Digitally cut vinyl stickers can also be used on a silkscreen frame for silkscreen printing: the process is the same, but the sticker can be removed and frame reused more often. It is also possible to use a roll of copper sticker with the vinyl cutter: in this way we can cut long and flexible electronic circuits, which can be glued onto many different surfaces.

SOFTWARE AND KNOWLEDGE REQUIRED

This is probably the easiest technology in a Fab Lab. Besides learning how to operate a vinyl cutter, this technology only requires knowledge of creating 2D vector designs, since it can only cut. These drawings can be created with many types of software, for example the proprietary software Adobe Illustrator or open source software Inkscape.

Small CNC milling

TECHNOLOGY

Historically, CNC milling machines were the first computer numerically controlled (CNC) manufacturing machines. This technology has developed in many directions in the past decades, and many different machines are available. This technology is also one of the most commonly used processes in industry: it is available in many factories, not just Fab Labs and makerspaces. Milling is the machining process of using rotary cutting tools with multiple cutting points that remove material by performing many separate small cuts. CNC milling machines enable the cutting tool and the piece of material to be moved or rotated along multiple axes by a computer. The most common CNC milling machines, especially in Fab Labs, have three axes: the rotary cutting tool is moved along X, Y and Z. Furthermore, four axes can be attained if the piece of material has one axis of rotation, and five if it has two axes of rotation. More axes of movement and rotation can be achieved with industrial robots, which have a completely different setup to other milling machines, even if the rotary cutting tool technology is the same. However, most of the time CNC milling machines in Fab Labs have three axes, and only sometimes four. Milling with three axes, we can produce pieces that are necessarily flat on one side (where the material is attached to the machine); milling with four axes, the pieces produced can be sculpted along 360°. There are many different designs and sizes for cutting tools, for different materials, geometries and effects: each Fab Lab typically only has a small subset of all the possible cutting tools. Fab Labs typically have small and desktop CNC milling machines that usually don't have the power or strength needed to mill hard metals, but can handle softer materials like aluminum, plastics, wood, wax or FR1 boards (layered boards of paper, glue and copper) for creating custom PCBs.

SOFTWARE AND KNOWLEDGE REQUIRED

CNC milling machines are probably the most difficult technology to operate in a Fab Lab, since there are many possible options, geometries and processes to consider when preparing a project. The project can be either a 2D CAD or vector drawing (created with proprietary software like AutoCAD, Rhinoceros 3D, freeware software like DraftSight, or open source like LibreCAD and FreeCAD, but also with proprietary software like Adobe Illustrator or open source software like Inkscape) or 3D model (created with proprietary software like AutoCAD, Rhinoceros 3D, SolidWorks, or open source software like Blender and FreeCAD). These 2D or 3D geometries then need to be transformed into the path for the cutting tool: CAM software is used for this purpose, where all the features of the job (tool size and shape, material properties, and size) need to be entered and processed before sending the data to the machine. PCBs can be produced with these machines, converting the layout to path movements in specific CAM software. Using small machines it is therefore possible to create small objects, PCBs or counter-molds, for later casting silicon and other resins (see the Molding and Casting section).

Big CNC milling

OBJECTS AND PROCESSES ↓

TECHNOLOGY

Besides small CNC milling machines, Fab Labs also have bigger CNC milling machines that are specifically designed for working with larger pieces of wood (or softer materials, like plastics). These machines are usually called CNC routers, and are not powerful or strong enough for milling metals, even if some of them can mill aluminum. CNC wood routers typically have three axes (but there are smaller ones with four axes as well), and their working area ranges from approximately 1.2 m (X) x 1.2 m (Y) x 0.15 m (Z), to 2.4 m (X) x 1.5 m (Y) x 0.15 m (Z).

These machines are based on the same technology as small CNC milling machines, but with bigger sizes and different details. As in small CNC milling machines, the piece of material needs to be attached to the machine and the position of the cutting tool needs to be calibrated and set before cutting away the material. These machines are usually not as precise as the small CNC milling machines, so more precise jobs like PCBs are very difficult to undertake with these machines (but the precision varies from machine to machine, and some machines can do very precise work). These machines are thus usually used for building furniture or bigger structures out of wood and other materials.

SOFTWARE AND KNOWLEDGE REQUIRED

CNC milling machines are probably the most difficult technology to operate in a Fab Lab, since there are many possible options, geometries and processes to consider when preparing a project. The project can be either a 2D CAD or vector drawing (created with proprietary software like Auto-CAD, Rhinoceros 3D, freeware software like DraftSight, or open source like LibreCAD and FreeCAD, but also with proprietary software like Adobe Illustrator or open source software like Inkscape) or 3D model (created with proprietary software like AutoCAD, Rhinoceros 3D, SolidWorks, or open source software like Blender and FreeCAD). These 2D or 3D geometries then need to be transformed into the path for the cutting tool: CAM software is then used for this purpose, and all the features of the job (tool size and shape, material properties and size) need to be entered and processed before sending the data to the machine.

Molding and casting

TECHNOLOGY

It is easy to use the technologies available in Fab Labs to create a single object; it is less easy to create small series of objects. The really critical element is usually time: with many users working in the lab, it is difficult for one person to use it for many days (unless the business model of the lab allows some users to rent the lab for several days). However, with enough time and space in the working area of each machine, it is possible to create more copies of one object. And there are also some processes, such as molding and casting, that directly enable users to create small series of objects.

Molding and casting are two separate but related processes. With molding, we create a mold from an existing object or a CNC milled block (usually machinable wax for greater precision and detail, but it can be done with other materials as well), also called a counter-mold. If we use an existing object for the creation of the mold we can use the mold for making multiple copies of existing objects. Alternatively, if we mill a new counter-mold and create a mold from it, we can create new objects. Molds are typically made with silicon or urethane resins, and some are even food-safe: we can then use the mold safely for the production of food.

With casting, we create one or more objects from a mold, using many kinds of plastic resins with different characteristics. Most of these resins can be also colored with specific pigments, creating colored objects. A release agent is typically sprayed onto the mold to make removal of the hardened resin from the mold easier. Some resins harden in a few minutes, but some silicon and resins take up to 16 hours, which can make this process very long. It is, however, a very easy manual process for creating small series of objects made from plastic or other materials like concrete. Resins, silicons and urethanes are normally created through the mixing of two different components: with some resins the process may lead to toxic vapors, and it is therefore important to create a specific space or setting for this process.

SOFTWARE AND KNOWLEDGE REQUIRED

Molding and casting processes require no specific knowledge or software, since they are both handwork processes. Knowledge about the specific materials and their reaction is not a prerequisite, but it is important to read the technical sheet provided with them.

3D printing: Fused Deposition Modeling (FDM)

TECHNOLOGY

Among all the digital fabrication technologies, 3D printing is the one currently receiving the most attention. 3D printing technologies, however, are rather a family of many different technologies, which have been developed since the 1980s; their speed and quality have since been improved, and their cost to the public lowered. These are the main reasons behind 3D printing's current popularity. Furthermore, more than other digital fabrication technologies, 3D printing technologies embody the idea of generating a physical object – almost out of thin air – from digital and immaterial geometry. Many 3D printing technologies are beginning to be able to produce finished goods and not just prototypes, but these technologies are typically very expensive: too much so, at least, for a normal Fab Lab. In conclusion, 3D printing technologies are becoming increasingly important in Fab Labs, but above all for prototyping and communicating what a Fab Lab can do (especially compared to the popularity of other technologies like laser cutting and CNC milling).

Among the many 3D printing technologies (which are often referred to as additive manufacturing, due to their ability to add layer over layer of material) the most popular in Fab Labs is Fused Deposition Modeling (FDM), also called Fused Filament Fabrication. FDM technology works by melting a plastic filament through a heated nozzle on a plane. Both plane and nozzle can be controlled by a computer, and moved about in order to build the desired geometry (each machine has a different design in terms of how much the nozzle and the plane can move, and how their movements are coordinated). The 3D printer melts the filament layer over layer, and each layer slowly cools down. Many materials are available, especially thermoplastics such as ABS, PLA, polycarbonate and polyamides: these are plastics that become moldable above a specific temperature and return to a solid state upon cooling. Their mechanical qualities are very good; however, the surface finishing can sometimes be rough, depending on the resolution and therefore on the speed of the process.

SOFTWARE AND KNOWLEDGE REQUIRED

Using 3D printers is a rather simple process: the difficulty resides in designing a 3D model that can be 3D printed successfully. The project needs to be a 3D model, created with proprietary software like AutoCAD, Rhinoceros 3D, SolidWorks, or open source software like Blender and FreeCAD. The 3D geometry needs to be processed in specific CAM software called the slicer, which prepares the geometry for the 3D printing process by slicing it into thin layers.

3D printing: Stereolithography (SLA)

OBJECTS AND PROCESSES ↓

TECHNOLOGY

FDM is now the most common 3D printing technology available in Fab Labs, thanks to the availability of low-cost and open source machines that can be designed and built in-house. But low-cost and open source Stereolithography (SLA) machines are starting to appear on the market, a clear sign that this faster but more expensive technology, with higher resolution and surface finishing, is going to spread to even more Fab Labs in the near future.

Stereolithography works in a similar way to FDM: the 3D geometry is sliced into thin layers in the computer, and then the 3D printer creates the object by solidifying these layers from a specific material. The difference here lies in how the material is solidified and which material is used: in Stereolithography, a UV laser or another similar power source cures a vat of liquid ultraviolet curable photopolymer resin. For each layer, the laser beam traces a cross-section of the part pattern on the surface of the liquid resin, polymerizing it and joining it to the layer below. After each layer, the elevator platform that holds the solidified object descends a distance equal to the thickness of a single layer. A resin-filled blade then sweeps across the surface, re-coating it with fresh material. On this new liquid surface, the laser traces the new section, adding another layer of solid resin to the object. When the object is finished and taken out of the 3D printer, it is usually immersed in a chemical bath to clean off excess resin. After this, it is cured in an ultraviolet oven. Like FDM, this technology requires the use of supporting structures which support and attach the object to the elevator platform: these structures are calculated and designed by the CAM software, and have to be removed manually at the end of the process.

SOFTWARE AND KNOWLEDGE REQUIRED

Using 3D printers is a rather simple process: the difficulty resides in designing a 3D model that can be 3D printed successfully. The project needs to be a 3D model, created with proprietary software like AutoCAD, Rhinoceros 3D, Solid-Works or open source software like Blender and FreeCAD. The 3D geometry needs to be processed in specific CAM software called a slicer, which prepares the geometry for the 3D printing process by slicing it into thin layers.

3D printing: Inkjet Powder Printing (3DP)

OBJECTS AND PROCESSES ↓

TECHNOLOGY

3D printing technologies are becoming an increasingly important feature in Fab Labs: FDM machines are very popular, while SLA machines are entering the scene with recent low-cost and open source versions. There exist other technologies that are either too new or too expensive to form part of the Fab Lab inventory. But at least one other technology has been observed in some Fab Labs, and a few low-cost and open source versions are emerging. Powder-based and similar to traditional inkjet printing, Inkjet Powder Printing is probably the only technology that could strictly be called 3D printing, in the sense of, 2D printing lifted to the third dimension.

Like FDM and SLA, the 3D geometry is sliced into thin layers, but here each layer is solidified by liquid glue drops deposited by the inkjet print head. In some machines, the inkjet head also deposits a colored ink, rendering the object fully colored. After each layer is finished, a thin layer of powder is spread across the completed section and the process is repeated: each layer is attached to the previous one with the same glue. When the object is complete, the superfluous powder is manually removed and may be reused. Then the object is solidified further by infiltrating it with a specific glue or epoxy. This technology is faster than FDM and SLA, since, because loose powder supports overhanging features and stacked or suspended objects, it does not need any additional support structure. Surface finishing is quite good, but not up to the standard of SLA, while its mechanical qualities are good, but likewise inferior to those of SLA.

SOFTWARE AND KNOWLEDGE REQUIRED

Using 3D printers is a rather simple process: the difficulty resides in designing a 3D model that can be 3D printed successfully. The project needs to be a 3D model, created with proprietary software like AutoCAD, Rhinoceros 3D, SolidWorks, or open source software like Blender and FreeCAD. The 3D geometry needs to be processed in specific CAM software called a slicer, which prepares the geometry for the 3D printing process by slicing it into thin layers.

Thermoforming

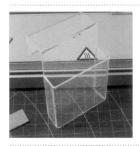

TECHNOLOGY

Thermoforming is a manufacturing process that uses heat to transform a wood or plastic sheet – such as PVC or PMMA – with the shape of a mold. Few Fab Labs have thermoforming machines, but, as they are relatively cheap and easy to build from scratch, their availability is increasing. In Fab Labs, molds are usually either CNC milled or 3D printed: thermoforming machines available in Fab Labs are not generally digital fabrication technologies, but are used in combination with other digital fabrication technologies available in the labs.

SOFTWARE AND KNOWLEDGE REQUIRED

Thermoforming processes require no specific knowledge or software, since they are all handwork processes.

3D scanning

OBJECTS AND PROCESSES   ↓

TECHNOLOGY

A 3D scanner analyzes a real-world object or environment by collecting data on its shape and sometimes even its color, and by transforming the data into a 3D geometry. There are many different technologies available for 3D scanning. Few Fab Labs have the more expensive ones, and most of them have the cheapest ones. A digital geometry reconstructed from real-world data can be a faster and more accurate way of creating a 3D model: it can be used in movies, video games, industrial design, prosthetics, reverse engineering, or with cultural artifacts.

Some 3D scanners can work through the direct contact of the device with the object: a few CNC milling machines even have 3D scanning abilities. Many other 3D scanners work without a direct contact with the object: for example, laser scanners create a 3D image through the triangulation of a laser dot or line, while structured light scanners measure the deformation of a projection of a pattern of light on the object. It is also possible to create a 3D scan of an object or an environment using software that analyzes a series of pictures. Some of this software is accessible for free, making it one of the most widely available options in Fab Labs. Otherwise, a very common low-cost device that is used for 3D scanning in Fab Labs is the Microsoft Kinect: a device for motion sensing input that was originally developed for controlling video games, and then applied to computers. By using an infrared projector, a camera and a special microchip, the Kinect can track the movement of objects and individuals in three dimensions, and therefore also their geometry; it has lower resolution than other technologies, however.

SOFTWARE AND KNOWLEDGE REQUIRED

3D scanning is a rather simple process: depending on the specific technology adopted, there will exist related custom software for reading the data from the 3D scanner or for transforming the pictures into a 3D model. The resulting 3D geometry can be cleaned and fixed with proprietary software like AutoCAD, Rhinoceros 3D, SolidWorks or open source software like Blender and FreeCAD.

Digital embroidery

TECHNOLOGY

In Fab Labs, many users work with textiles and leather, creating simple clothes or wearable computing experiments. These projects can be carried out by hand, with traditional sewing or embroidery machines, or with digital embroidery machines, which are becoming very popular. Digital embroidery machines have a framing system that holds the framed area of fabric under the sewing needle and moves it automatically, creating a design from a digital file. These machines can have one needle and one thread (which can be changed for other threads after each task, mixing different colors and materials, and even for conductive threads for creating wearable computing projects) or multiple needles and multiple threads (which are ready to be used automatically by the machine).

SOFTWARE AND KNOWLEDGE REQUIRED

Digital embroidery machines function in the same way as traditional embroidery machines, albeit with a digital control: it is therefore important to learn the basics of sewing and embroidery in order to use them properly. Digital patterns need to be designed with specific software provided by the manufacturer or generic software like Embird, which can be bought or downloaded. The software acts like a CAM, converting a raster image to the pattern that will be created, and simulating it digitally.

Handwork

TECHNOLOGY

An important aspect of digital fabrication technologies is that, more often than not, they require further handwork. Objects may not be cut or milled precisely: there could be imperfections or surfaces that need to be smoothed, for instance. Fab Labs therefore always have a wide range of hand tools, which vary from lab to lab, from drills to vertical drills, handsaws, circular saws, jigsaws, hammers, rasps, files, scissors, Stanley knives, hot glue guns, spray paint, sandpapers, screwdrivers, wrenches, and so on.

Peter Troxler

What is a Fab Lab for?

The philosophy, general and specific purposes

Fab Labs are an important part of the Maker ecosystem. The whole movement has emerged thanks to its ability to network projects, people and places. But each project, business, laboratory and community in the Maker movement has its own perspectives, philosophies and practices. This chapter critically addresses this issue, laying out the most common approaches to be found among the different laboratories and dynamics of the Fab Lab network and the Maker movement as a whole.

Structure, ideas and challenges of the Fab Lab

 ↓

Obsolescence:
competition from the diffusion
of cheap digital fabrication
technologies   Ⓐ

Governance: Ⓑ
development of
the concept and
the evolution of its
local modifications   **FAB LAB CONCEPT** ①

Competition: Ⓒ
new/other models
of community
workshops

1. Digital and personal fabrication
2. Everybody can be a designer
3. Technology / STEM education
4. Maker movement
5. Third Industrial Revolution

Ⓓ

Balance:
local communities
within the global
community

Ⓔ

Technology:
it is not neutral
and societal,
cultural political
dimensions should
be addressed

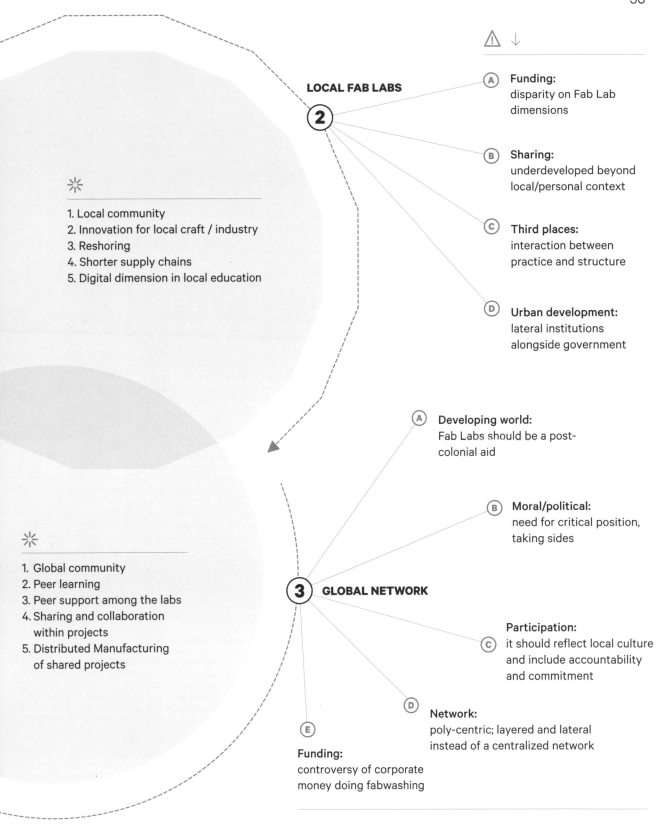

⚠ ↓

LOCAL FAB LABS

2

Ⓐ **Funding:** disparity on Fab Lab dimensions

Ⓑ **Sharing:** underdeveloped beyond local/personal context

Ⓒ **Third places:** interaction between practice and structure

Ⓓ **Urban development:** lateral institutions alongside government

※
1. Local community
2. Innovation for local craft / industry
3. Reshoring
4. Shorter supply chains
5. Digital dimension in local education

Ⓐ **Developing world:** Fab Labs should be a post-colonial aid

Ⓑ **Moral/political:** need for critical position, taking sides

3 **GLOBAL NETWORK**

Ⓒ **Participation:** it should reflect local culture and include accountability and commitment

Ⓓ **Network:** poly-centric; layered and lateral instead of a centralized network

Ⓔ **Funding:** controversy of corporate money doing fabwashing

※
1. Global community
2. Peer learning
3. Peer support among the labs
4. Sharing and collaboration within projects
5. Distributed Manufacturing of shared projects

The philosophy
of Fab Labs

The question asked most often during the early days of Fab Labs was: 'What, exactly, is a Fab Lab?' As we have seen, Fab Labs have since become well-known as a phenomenon, and the question has now become: 'What, exactly, is a Fab Lab for?' This chapter aims to answer this question on three levels – by describing the overall philosophy of Fab Labs, by describing the general purpose of Fab Labs, and by presenting various insights into what specific purposes Fab Labs serve, and what their impacts are. Given that the concept of the Fab Lab is constantly evolving, the chapter concludes with a consideration of the challenges Fab Labs are likely to face in the future.

The initial purpose of a Fab Lab was twofold: to serve as an educational environment for engineers, and to be a manifestation of the first phase of an ambitious research program – the Center for Bits and Atoms – that would eventually shed light on the question of how bits and atoms can be considered to be essentially the same thing[1]. That program also included two kinds of outreach activity, one in education and one in development. In education, it sought to engage children in science using a constructionist approach, to develop lightweight programmable modules that could interact in large-scale physical systems, and to develop educational material that would complement traditional curricula. In development, the CBA envisioned bringing technology from the laboratory to community-based groups, so they could solve local problems of healthcare monitoring, regional communications, environmentally-friendly food production and global connections for electronic commerce[2]. The course titled How To Make (Almost) Anything, the root of Fab Lab principles, proved initial assumptions wrong. Unexpectedly, it was oversubscribed, and the best students, according to Gershenfeld's own accounts[3], were not engineers but artists. The initial idea of the Center for Bits and Atoms was to realize a handful of outreach 'field' Fab Labs. Yet again, the reality was different, as demand for Fab Labs was growing, and the CBA had to adapt to and deal with that demand. One solution was state-funded programs, developed in partnership with CBA. However, in some countries, such as the Netherlands, people increasingly started to set up Fab Labs without the direct intervention of Gershenfeld. The initial concept of a one-size-fits-all Fab Lab – sticking to the equipment and consumables suggested by the CBA – was challenged when the Amersfoort Fab Lab was set up 'in 7 days with 4 people and about €5000'[4]. This approach was later integrated into the Fab Lab philosophy[5] as part of the narrative of '10k, 100k, 1000k labs'. The basic philosophy of Fab Labs has developed over time. Predictions have been superseded faster than anyone expected. Leaders in the Fab Lab community were relatively conservative in their projections, and grappled with adjusting them to a reality that repeatedly developed more quickly and in different directions than had been expected. They nevertheless remained open to these developments and – although they sometimes struggled – were willing and able to combine them with the principles set out at the beginning of the program, in an expansive rather than a reductive manner.

1• MIT News Office, 2001. 'Media Lab creates Center for Bits and Atoms with NSF grant'. MIT News. Available at: http://newsoffice.mit.edu/2001/bits-1128

2• Gershenfeld, N., 2001. Center for Bits and Atoms. Project Summary. ITR/SY Grant Proposal.

3• Gershenfeld, N., 2005. FAB: The Coming Revolution on Your Desktop – From Personal Computers to Personal Fabrication, Basic Books.

4• Zijp, H.G., 2011. The Grassfoots FabLab Instructable – Or How To Set Up a FabLab in 7 Days with 4 People and About €5000. Available at: http://www.fablabamersfoort.nl/downloads/fablab-instructable.pdf

5• Editor's note: for a longer explanation of the '10k, 100k, 1000k labs' range, see Chapter 4 of this book.

For Gershenfeld, the general purpose of Fab Labs is to make what you cannot buy in shops[6]. The killer app for personal fabrication in the developed world is technology for a market of one: personal expression in technology that touches a passion unlike anything I've seen in technology for a very long time. And the killer app for the rest of the planet is the instrumentation and the fabrication divide: people locally developing solutions to local problems.[7] This purpose is echoed in the Fab Charter[8], which states that labs are 'enabling invention by providing access to tools for digital fabrication'. This inventiveness is a general quality in the Maker movement, of which Fab Lab is a part. Particularly in the developed world, the individual products of Makers need not be – and often are not – useful, or even beautiful, to anyone other than the individual Maker, and many might even be considered wild, crazy or weird, as some of the more extravagant exhibits at Maker Faires demonstrate. Fab Labs – and more generally the Maker movement – in the developed world can be read in at least four ways, depending on context and critical perspective.

ONE: Fab Lab and the Maker movement can be read as a mainly bourgeois pastime, a token act of rebellion that at heart is just a new form of entertainment and consumption. Only very few members of the movement develop fundamentally new things; the vast majority copy existing projects, adding small and mostly cosmetic adaptations. The genealogy of the Rep Rap[9] project and its countless clones are a case in point. The success of kits, the popularity of Thingiverse[10] and Instructables[11] to share and find projects, and the steadily growing number of visitors to the Maker Faires constitute further telling evidence, as does the readership demography of Make Magazine: eight in ten Makers are male, with a median age of 44, report a high median household income of $106,000, and are married home-owners with children under 17. 97% attended college, four in ten hold post-graduate degrees, and 83% of Makers are employed.[12]

TWO: Fab Lab and the Maker movement can be read as an innovation in technology education. They resonate with industry bodies in many countries that fear a decline in the technically skilled workforce. However, critics have accused industry of manipulating the labor market, 'inflating supply and depressing demand for scientists and engineers'[13]. And research has shown that the problem of a diminishing technical workforce 'is one of location mismatch: talented people are available but not always in the places where they are needed'[14]. Nevertheless, many governments subsidize STEM education. The educational method that corresponds best to a Fab Lab environment is 'learning-by-making'[15], constructionist learning, as opposed to traditional instructionist pipeline models of transmitting knowledge. Beyond being an educational method, constructionist learning also has epistemological implications. It is concerned with the nature of knowledge and knowing: what counts as knowledge? How is this knowledge structured? It challenges the canonical epistemology of STEM education – that knowledge is abstract, impersonal and detached – and counters it with epistemological pluralism.[16]

The general purpose of Fab Labs

6• Gershenfeld, N., 2012. How to Make Almost Anything – The Digital Fabrication Revolution. Foreign Affairs, (November/ December). Available at: http://www.foreignaffairs.com/articles/138154/neil-gershenfeld/how-to-make-almost-anything

7• Gershenfeld, N. (2006). Unleash Your Creativity in a Fab Lab. TED Talk. Available at: http://www.ted.com/talks/neil_gershenfeld_on_fab_labs

8• CBA, 2012. The Fab Charter. Center for Bits and Atoms. Available at: http://fab.cba.mit.edu/about/charter/

9• Anon, RepRap - RepRapWiki. Available at: http://reprap.org/

10• Anon, Thingiverse - Digital Designs for Physical Objects. Available at: http://www.thingiverse.com/

11• Anon, Instructables - DIY How To Make Instructions. Available at: http://www.instructables.com/index

12• Make/Intel (2012). An In-depth Profile of Makers at the Forefront of Hardware Innovation. Maker Market Study and Media Report. Available at http://cdn.makezine.com/make/sales/Maker-Market-Study.pdf

13• Macilwain, C., 2013. Driving students into science is a fool's errand. Nature, 497(7449), pp. 289–289.

14• Craig, E. et al (2011). No Shortage of Talent: How the Global Market is Producing the STEM Skills Needed for Growth. Research Report. Accenture Insititute for High Performance.

15• Papert, S. and Harel, I. (1991). Situating Contstructionism. In: Papert, S. and Harel, I. (eds.) Constructionism. New York, NY: Ablex Publishing Corporation, pp. 1-11.

16• Turkle, S. and Papert, S. (1991). Epistemological Pluralism and the Revaluation of the Conrete. In: Papert, S. and Harel, I. (eds.) Constructionism. New York, NY: Ablex Publishing Corporation, pp. 161-191.

THREE: Fab Lab and the Maker movement can be read as a new renaissance that is reuniting the liberal arts with science and engineering in a contemporary and playful way. This notion of play is expressed both in the products and artifacts of the Maker movement, and in the constructionist approach to learning discussed above. One aspect of play is to try different approaches to a situation or problem and learn from the success or failure of these approaches. Another, complementary aspect of play is that this trial-and-error approach is not impeded by a fear of failure. Failing and learning from failure are important and encouraged, particularly when failure is quick and cheap. In engineering, this means a step back from rigid, multi-disciplinary, time-consuming systems engineering approaches, and the adoption of a highly iterative, interdisciplinary and rapid mode of working. Airbus has implemented this approach in their internal Protospace, where they are able to develop new subsystems for aircraft within weeks rather than the industry standard of several years. Such an approach is much more fundamental than just 'design thinking', given that the result of the process is not just a mock-up, but also a fully functional, complex product. With respect to the arts, artists have engaged with technology and science for a long time now; however, art theory had a tendency to put 'art and technology', 'media art', 'computer art', 'Internet art', 'art and science', etc. into separate pigeonholes, which makes it harder to gauge the overall contribution of the arts. The umbrella term 'hybrid art' is increasingly used to indicate how artists are doing research and technology development that would be rejected by mainstream science and industry, despite being of critical societal relevance.

17• Anderson, C., 2012. Makers: The New Industrial Revolution. Crown Business.

FOUR: Fab Lab and the Maker movement can be read as a 'new industrial revolution'[17]. This revolution has a number of ingredients: empowerment through mastery of technology gives people the means to understand and build seemingly very complex things. It also allows people to understand that the way technology works is in most cases not a technological given but determined by decisions made by humans – often engineers working for big corporations whose motives might not always be to build socially useful products – and therefore expose corporate strategies, among them design for obsolescence, the notion that you don't own a product if you can't open it, and the quest to repair broken goods, or the gendered design of technology. Another ingredient of the revolution is the move away from globalized mass production to local, small-batch production and lateral and networked forms of organization[18]; indeed the Fab Charter also states that 'Fab Labs are a global network of local labs'[19]. Technical empowerment and local production give rise to the hope that this revolution will create new work and income, in particular for the high number of unemployed young people, in an emerging collaborative and sharing economy. In its contemporary manifestation, however, the sharing economy has slid rapidly 'from neighborliness to the most precarious of casual labor'[20]. A final revolutionary aspect of Fab Lab and the Maker movement is their impact on scientific endeavor. Technical empowerment allows individuals to participate in and carry out scientific research. Citizen science allows for large scale, distributed and long-

18• Rifkin, J., 2011. The Third Industrial Revolution: How Lateral Power is Transforming Energy, the Economy, and the World. Palgrave Macmillan.

19• CBA, 2012.

20• Slee, T. (2014). Sharing and Caring. Jacobin, 13. Available at: http://www.jacobinmag.com/2014/01/sharing-and-caring/

term data collection and investigation, greatly expanding the capacities of hybrid art mentioned above. It is bound to complement and contrast with established production systems of scientific insight.

Outside the developed world, it has been suggested that Fab Labs are less about the 'technology for a market of one', and more about solutions to local problems. A recent World Bank showcase of projects[21] comprised a spirometer to measure breathing volumes, an optical microscope built from paper with a cheap lens, a manual blood centrifuge, 3D printed prosthetics, water and air quality sensors, an algorithm that enables simple webcams to monitor the heart and breathing rate of newborn babies or the swaying of buildings, a 3D printer built from broken computers, plastic recycling for 3D printing, drones, a traffic counter and street furniture. Yet only one third of these projects actually originated from developing countries (the blood centrifuge, the 3D printer, the air quality sensor and the public space furniture project). The other nine projects were located in the United States and in Spain. The World Bank's involvement in digital fabrication is at a very early stage, and the selection of projects is not meant to be representative, but to illustrate to a Western audience how digital fabrication would enable people to build very useful equipment, at a fraction of what it would cost to buy it from traditional manufacturers. The selection of projects does, however, illustrate the challenges and dilemmas development aid faces, even at a grassroots level.

The challenges are multiple. There is certainly a genuine desire to help developing countries to overcome what is perceived, from a Western point of view, as poverty. There is the almost magical potency of technology when it is utilized by ingenious young people to create intricate devices: empowerment through technology. The challenges create interesting contrasts: spirometer versus street furniture, computer algorithms versus 3D printers. Western aid often assumes that technology, materials, energy and knowledge can easily be made available. Yet each of these four resources comes with its own constraints. Importing certain technology is on occasion cumbersome; recreating it on the spot is hard. There are plenty of materials available in developing countries, but they are substantially different in quality to the ready-to-use staples available in the developed world. Energy supply, which is taken for granted in the West, is intermittent and might need to be prioritized, as does Internet access. And knowledge, finally, takes on a completely different shape outside rationalized Western cultures; what is supplied as knowledge might only have limited resonance with what is considered knowledge in the local context. Yet more challenges arise from the way aid is delivered and Fab Labs are set up in the developing world. It is indeed true that the principle of setting up Fab Labs is radically different to traditional aid, which is channeled through large aid organizations and the bureaucracies of the receiving countries, delivering food, clothing and housing as commodities, reinforcing dependencies and wasting large parts of aid money along the way. Fab Labs work in a much more direct way, and are supposed to be means to

21• Muente-Kunigami, A. et al., 2014. Makers for Development. Showcasing the Potential of Makers, World Bank - USAID - Fab Foundation - Fab Lab BCN. Available at: http://www.usaid.gov/GlobalDevLab/makers4development.

survival, rather than an end in themselves. Yet there is a huge difference between a Fab Lab established at a university, in collaboration with a Western counterpart, and a grassroots Fab Lab developed for and with the people. Free access, prioritized in the Fab Charter, can hardly be granted at a Fab Lab situated inside the highly guarded campus of a university in the developing world, and that is only one issue. Access to knowledge, international communication, energy and materials is another. There is also the assumption that aid in the form of technology, be it simple computers or a complete Fab Lab infrastructure, is neutral: upon critical reflection, this assumption appears redundant, as technology must be understood as an embodiment not only of technological functions but also of cultural and societal practices and routines, which might give rise to wariness regarding neo-colonialism, more than the liberation and emancipation of the developing world. Lastly, the term 'Developing World' is a shameless generalization; diversity among – and even within – countries is enormous, so it is hardly conceivable that a one-size-fits-all approach could meaningfully satisfy the aims of development aid. The World Bank has shown[22] that 'the naive application of complex contextual concepts like participation, social capital, and empowerment is endemic among project implementers and contributes to poor design and implementation'. The challenge for the Fab Lab community is to turn the Fab Lab approach into an instrument of post-colonial aid that takes into account local heterogeneity and the historical, political and social environment in which it is supposed to be implemented, so that individuals and groups can become authors of their own development.

22• Mansuri, G. & Rao, V., 2004. 'Community-Based and -Driven Development: A Critical Review'. The World Bank Research Observer, 19(1), pp.1–39.

Insights into the specific purpose of Fab Labs

23• Anon, Fab Lab Wiki - by NMÍ Kvikan. Available at: http://wiki.fablab.is/wiki/Main_Page

24• Anon, Fab Lab Wiki - by NMÍ Kvikan. Available at: http://wiki.fablab.is/wiki/Main_Page

25• Troxler, P., 2010. Commons-Based Peer-Production of Physical Goods: Is There Room for a Hybrid Innovation Ecology?, Rochester, NY: Social Science Research Network. Available at: http://papers.ssrn.com/abstract=1692617

26• Stelzer, R. & Jafarmadar, K., 2013. Low-Threshold Access to Fab Labs through Training Programs and Outreach Activities. In Fab 9 Research Stream. Yokohama. Available at: http://www.fablabinternational.org/fab-lab-research/proceedings-of-the-fab-9-research-stream.

In terms of statistical evidence, Fab Labs are notoriously elusive, even more so when it comes to demonstrating the output, outcomes and impact of a Fab Lab, or the regional or global Fab Lab networks. The sources for quantitative information on the Fab Lab movement – the lab lists at fablabs.io [23] and wiki.fablab.is[24], which both come with their own peculiarities in terms of data quality – are able to demonstrate that the number of labs has been growing exponentially over the past decade, doubling every twelve to eighteen months. One can assume that the number of people reached by the Fab Lab movement has seen an equal development of rapid exponential growth.

However, this poses a problem for the validity of any study carried out at a certain point in time. For instance, when we surveyed Fab Lab business models in 2010, we found that one in five labs was hosted by an educational institution.[25] Re-checking that proportion in 2013 revealed a three in five ratio. Growth in the network is not homogeneous in many other respects: for example, whole countries often develop a relatively dense network over a period of only a few years, or participation in Fab Academy – one place where prospective lab staff can get training – has been roughly constant over some years, suggesting a linear rather than exponential growth of labs run by Fab Academy trained gurus. Very often the data available from individual Fab Labs, such as the report provided by Happylab in Vienna[26], mainly describe input variables – the number of members, the conversion rates of training, etc. – and give little account of members' reasons for starting to participate. The Happylab study assumes that members – mainly former students in their twenties – were on the whole familiar with digital design and wanted to use the machinery to realize their ideas. Of similar quality are the numerous surveys carried out by agencies or undergraduates in preparation for writing a business plan for a soon-to-be-opened lab. Fabien Eychenne found five archetypes of Fab Lab projects – proof of concept prototype, small series, collective projects in collaboration between different labs, unique artistic or arts student projects, and projects serving a niche market[27]. The collective projects are a particularly interesting case, as the common capabilities of Fab Labs are supposed to allow projects to be shared [28], and because an intuitive understanding suggests that DIY amateurs participate actively in cooperative projects[29]. In an in-depth study of seventeen collaborative projects, we explored whether and how knowledge is shared globally in the Fab Lab community. However, we found that sharing was far from the norm, and remained confined to local and personal networks [30]. This confirms the findings of a pair of ethnographic studies focusing on single labs, namely Maldini's[31] and Ghalim's[32] studies of the users of Fablab Amsterdam. Both studies put the claims of an imminent third industrial revolution in perspective. Maldini found that users of Fablab Amsterdam were mostly professionals, not amateurs. They produced mainly prototypes and mock-ups, disposable objects, personalized artifacts, small series or reproductions of rare pieces. And they mainly worked on their own, hardly sharing or collaborating on projects, although they felt part of this facility and its related community. The cases Ghalim reports confirm that picture of personal fabrication in the form of relatively disconnected, solitary projects or projects carried out in small, local groups of friends.

Camille Bosqué visited several Fab Labs and similar collective workshops in France, the US (Oakland, San Francisco and San Diego), Norway, the Netherlands and Japan, and found them to be 'places where design practices go on mainly without designers, on the fringes of industry and traditional production patterns'[33]. She found a creative field in which participants are not only acting upon an open technical object, but redefining social and political relations[34]. Sabine Hielscher studied 85 UK Fab Labs, hackerspaces and makerspaces, and in particular looked into the question of how these spaces presented themselves online with regard to these political and societal questions. She found, in contrast to Bosqué, that 'only a small number of workshops might directly explore and reflect upon how these activities sit within wider societal, environmental and cultural developments'[35]. While the difference in findings could stem from various differences between the two studies – the countries surveyed, the selection process for cases, the method of analysis (interviews versus document analysis), or the type of conversation (private or public) – the two studies document the diversity of positions Fab Labs take vis-à-vis societal and political questions, and the varying importance these questions have for individual labs. There are other types of labs, particularly those who identify

27• Eychenne, F., 2012. Fab Labs Tour d'horizon, Paris: Fing. Available at: http://fing.org/?Tour-d-horizon-des-Fab-Labs,866&lang=fr.

28• CBA, 2012.

29• Kuznetsov, S. & Paulos, E., 2010. Rise of the Expert Amateur: DIY Projects, Communities, and Cultures. In Proceedings of the 6th Nordic Conference on Human-Computer Interaction: Extending Boundaries. NordiCHI '10. New York, NY, USA: ACM, pp. 295–304. Available at: http://www.staceyk.org/hci/KuznetsovDIY.pdf

30• Wolf, P. et al., 2014. Sharing is Sparing: Open Knowledge Sharing in Fab Labs. Journal of Peer Production, (5). Available at: http://peerproduction.net/issues/issue-5-shared-machine-shops/

31• Maldini, I., 2014. Digital makers: an ethnographic study of the FabLab Amsterdam users. In A Matter of Design: Making Society through Science and Technology. Available at: http://www.stsitalia.org/conferences/ocs/index.php/STSIC/AMD/paper/view/58

32• Ghalim, A., 2013. Fabbing Practices: An Ethnography in Fab Lab Amsterdam. Master's Thesis. Amsterdam: Universiteit van Amsterdam (New Media and Culture Studies). Available at: http://www.scribd.com/doc/127598717/FABBING-PRACTICES-AN-ETHNOGRAPHY-IN-FAB-LAB-AMSTERDAM.

33• Editor's note: Camille Bosqué wrote about some of these Fab Labs in Chapter 1 of this book.

34• Bosqué, C., 2013. Hack/make: designing and fabrication in "labs" and collective workshops. In Fab 9 Research Stream. Yokohama. Available at: http://www.fablabinternational.org/fab-lab-research/proceedings-of-the-fab-9-research-stream

35• Hielscher, S., 2014. Examining web-based materials. A snapshot of UK Fab Labs and Hackerspaces. Grassroots Innovation Research Briefing 24. Available at: https://cied.ac.uk/webteam/gateway/file.php?name=a-snapshot-of-uk-fablabs-and-hacker-spaces.pdf&site=440

themselves as hacklabs, that explicitly form part of the broader anarchist/autonomist scene; hackerspaces, originally linked to the hacker subculture, are less overtly political and of a more liberal or libertarian orientation.[36] Equally, makerspaces do not normally affiliate themselves with a political identity, and while some are concerned that Maker Media might at a given moment monopolize the term, they are not too worried about being associated with the fake bourgeois revolution of the Maker Media ethos.

When it comes to educational initiatives, the landscape appears similarly fragmented and disconnected. Maker Media, partly with the support of DARPA, has been actively promoting the Maker approach to schools and educators since 2012, with books, kits and a whole Maker education initiative[37], including a network of young Makers' clubs across the US.[38] Paulo Blikstein's work at Stanford's Transforming Learning Technology Lab is less about propaganda and more academically rooted, a fact that is underlined by the FabLearn conferences. Blikstein started his educational program Fab@School in 2009[39], claiming it was the first program designed from the ground up specifically to serve grades 6–12, and reaching from Stanford to Palo Alto, Moscow, and Bangkok. Within the Fab Lab network, the role of Fab Labs in education started to become part of the discussion of Fab Lab operations at the annual gatherings around the same time[40] and grew into the international network FabEd, supported by the Fab Foundation and the US-based Teaching Institute for Excellence in STEM (TIES)[41]. This initiative has a strong focus on curriculum integration and development, and on student assessment. There are other initiatives, such as Gary Stager's and Sylvia Martinez's Invent to Learn[42], Emily Pilloton's Project H[43], Per-Ivar Kloen's and Arjan van der Meij's FABklas in The Hague[44], or the Hakidemia network, with its outreach activities in Eastern Europe and Africa[45]. Their common line of argument is to repeat the industry claim that the technically trained workforce is dwindling. As governments around the world bought into this argument, there emerged a market of public funding and corporate sponsorship available to sustain such activities. The availability of public money probably adds to the fragmentation of educational activities related to making. And this is probably the reason why every individual Fab Lab is involved in some way in primary, secondary or higher education, whether it is housed within a school or university or through outreach, increasing fragmentation even more. It is doubtful if such fragmentation allows labs to provide the best services to students and education, as experiences beyond success stories are not often shared or reflected upon by peers and educationalists.

When it comes to economic and business impact, the second most important activity after education is probably the installation of new labs. The notion in the Fab Charter that the network provides 'operational, educational, technical, financial, and logistical assistance beyond what's available within one lab'[46] is an indicator for this market. Gauging the size of this market, however, is difficult, because consultants and Fab Labs that share their expertise do not share details of their business relationships (if the relationships are made public

36• Maxigas, 2012. Hacklabs and Hackerspaces. Tracing Two Genealogies. Journal of Peer Production, (2). Available at: http://peerproduction.net/issues/issue-2/.

37• Maker Education Initiative, Maker Education Initiative – Every Child a Maker. Available at: http://makered.org/

38• Young Makers, Young Makers. Available at: http://youngmakers.org/

39• Blikstein, P., FabLab@School - Digital Fabrication for Education. Available at: http://fablabatschool.org/

40• See the programme of FAB5 at CBA, FAB5. Available at: http://cba.mit.edu/events/09.08.FAB5/index.html

41• Teaching Institute for Excellence in STEM, STEM Solutions | TIES. Available at: http://www.tiesteach.org/solutions/

42• Martinez, S.L. & Stager, G.S., 2013. Invent To Learn: Making, Tinkering, and Engineering in the Classroom, Constructing Modern Knowledge Press.

43• Anon, Project H. Available at: http://www.projecthdesign.org/

44• Anon, [FABklas] - Wat maken de makers van morgen? Available at: http://fabklas.nl/

45• Anon, HacKIDemia. Available at: http://www.hackidemia.com/

46• CBA, 2012.

at all), nor as a rule do new labs publicize their start-up costs. So proxies would have to be used, such as the number of new labs per year, the number of Fab Academy graduates, the larger grants announced and the vacancies for Fab Lab gurus. More problematic, however, is the suspicion that setting-up Fab Labs as a core source of income for existing labs could create a Fab Lab bubble that would collapse once the market was saturated, or that it could be rightly or wrongly exposed as a pyramid scheme, and discourage initiatives from associating themselves with the Fab Lab network.

The cumulative value Fab Labs create as a community resource, by offering open access for individuals and by running programs, is even harder to gauge and express. As it includes recreational, educational, social and societal components – Fab Labs are places to have fun, to learn, to interact and to build communities – this value is multi-dimensional. The process of creating value in that sense is rather complex, multilateral and often reciprocal[47].

Individual business models thus cannot just take the simplistic form of an abstract description of how a lab makes money by buying and selling goods and services, or 'what it offers, to whom it offers this and how it can accomplish this.'[48] While it is crucial to get the monetary side of running a Fab Lab right – the determination to succeed as a business – the blend of monetary and non-monetary streams of value creation are what makes a lab successful.[49]

Fab Labs – and the Maker movement – have been receiving growing interest as a tool for revitalizing the economy in countries (such as the US) and cities alike. For instance, the city of Barcelona has been promoting itself as Fab City ever since Vicente Guallart, director of the IAAC, which hosts the Barcelona Fab Lab, became the city architect under mayor Xavier Trias. The challenge to convert Barcelona into a 'Data-In-Data-Out' city – instead of one that imports products and exports trash – was apparently posed to Guallart and Trias by Neil Gershenfeld.[50] Guallart situates the Fab Lab at city-block level – an area of 10,000 m^2 and 1000 inhabitants – whereas official Fab City publications[51] typically talk about one Fab Lab per neighborhood (10,000–100,000 inhabitants, in Guallart's typology). The house bill to grant the Fab Foundation US federal charter status was based on one Fab Lab per 700,000 inhabitants.[52]

Notwithstanding this substantial discrepancy in numbers, which would at least suggest that there is little science behind such projections, there is a growing desire in cities to boast a Fab Lab. One is the often-cited inner city Boston Fab Lab[53], the first to be created out of MIT. It was featured in the first volume of Make magazine and there are plenty of anecdotes linked to the lab – that group of girls who 'set up the laser cutter on a street corner and held a high-tech draft sale, making things on demand from scrap materials, [and] also made $100 in an afternoon, a life-changing experience in a community with limited economic opportunities for them';[54] or Makeda Stephenson, who at the age of 13 built her own flight simulator, 'that

47• Troxler, P., 2013. Can We Think Differently of Fab Lab Business Models? In Fab 9 Research Stream. Yokohama.

48• Osterwalder, A., 2004. The business model ontology: A proposition in a design science approach. Lausanne: Université de Lausanne (Ecole des Hautes Etudes Commerciales). Available at: http://www.stanford.edu/group/mse278/cgi-bin/wordpress/wp-content/uploads/2010/01/TheBusiness-Model-Ontology.pdf

49• Editors' note: for a further explanation of this topic see Chapter 4 of this book.

50• Guallart, V., 2014. The Self-Sufficient City: Internet has changed our lives but it hasn't changed our cities, yet. New York: ActarD Inc.

51• Diez, T., 2011. Barcelona 5.0. Production, talent & networks, regenerating the cities of the future. Available at: http://cba.mit.edu/events/11.08.FAB7/Tomas.pdf and Alvarellos, S., 2012. 4/4 Fabbing & cities: Barcelona Fab City. complexitys. Available at: http://complexitys.com/english/44-fabbing-cities-barcelona-fab-city/

52• National Fab Lab Network Act of 2013. H.R. 1289, 113th Cong. (2013). Available at: https://beta.congress.gov/bill/113th-congress/house-bill/1289. The corresponding senate bill had the number dropped from the text. National Fab Lab Network Act of 2013, S. 1705, 113th Cong. (2013). Available at: https://beta.congress.gov/bill/113th-congress/senate-bill/1705

53• Jewell, M., 2005. Fab labs give kids a creative boost' Geared to spark imaginations through technology - Cecil Daily: Localnews. Cecil Whig. Available at: http://www.cecildaily.com/news/localnews/article_9735eb0e-a65a-5ca3-b699-677c218dbca3.html

54• Denison, C.C., 2005. Maker: Welcome to the Fab Lab. A Tour of Neil Gershenfeld's Fab Lab at MIT. MAKE: technology on your time, 1, p. 26.

Opposite page:
Feeding Plants, CODECZOMBIE,
digital clay and 3D SLS printing, 2013.

55• Jewell, M., 2005.

56• Gershenfeld, 2005, p. 27.

57• According to published grant details at
European Union Programmes, EU Grants
Successes. Available at: http://successes.
eugrants.org/

58• Holden, J., 2014. FabLab ethos puts
community first. New facility encourages
creativity, technology and social
inclusion. The Irish Times, June 16, 2014.
Available at: http://www.irishtimes.com/
business/fablab-ethos-puts-community-
first-1.1831623

59• ibid.

60• Osborn, S., 2012. Makers at Work. Folks
Reinventing the World One Object Or Idea at a
Time, Apress, p. 283.

10• Peek, G.J. and Troxler, P. (2014). City
in Transition. Urban Open Innovation
Environments as a Radical Innovation. In:
Schrenk, M., et al. (eds.) Real Corp 2014
Proceedings. Vienna: Corp. Available at:
http://programm.corp.at/cdrom2014/
papers2014/CORP2014_40.pdf

lets her "fly" an airplane of her design over an alien planet born of her imagination'[55]; or the girl who 'chose to make a diary security system that would take a picture of anyone (like her brother) who came near her diary'[56].

Another example are the two Fab Labs in Derry and Belfast in Northern Ireland. These were funded mainly by the European Union Peace Program with £1.35 million[57], of which initial capital costs were about €35,000[58]. They are supposed to contribute to reconciliation and peacebuilding by giving marginalized target groups in Northern Ireland access to the designing and making tools to explore and develop creative and entrepreneurial skills. According to their own account, they served over 100 people per week in each lab and helped to create 'a number of new businesses'[59]. It is still too early to assess the impact of the two Northern Irish Fab Labs, or to confirm that the massive funds poured into this venture indeed resulted in measurable results in terms of peace and reconciliation.

In other cities, the environment has been less favorable to Fab Labs, and labs with a budget of millions are rare. In Turin, for instance, the Fab Lab made its first appearance during an exhibition celebrating 150 years of the Italian nation, and depicting its bright future: in the case of the Fab Lab, the future of work. However, once the exhibition closed, none of the political entities of the city showed any interest in maintaining the project and supporting the community that grew up around it. It was down to local business owner Massimo Banzi of Arduino, who was initially tasked with setting up the 'exhibit', to keep it running[60], immediately sparking an explosive growth of Fab Labs and the Maker movement in Italy.

The development in Turin is more typical of how Fab Labs develop in cities: they emanate from an existing need that manifests itself in a certain community. The Fab Lab in Zurich was initiated by a group of designers and architects as a shared resource and a place where like-minded people could meet. The multiple labs in Rotterdam all stem from a specific local need – Stadslab is a university Fab Lab that specializes in sensors and Internet of Things; RDM makerspace is located in a business incubator setting; HET LAB caters for primary education; Made in 4 Havens establishes local production facilities for local designers; and Roffab brings mini-makerspaces to neighborhoods. Private initiatives – the Rotterdam Maakstad institute for industrial recreation and the Platform for Digital Manufacturing – were established to connect these labs with the wider industry.

Such developments are much more in line with modern views of urban development, in which new lateral institutions play an important role, alongside governments and large companies, in facilitating distributed and collaborative action.[61]

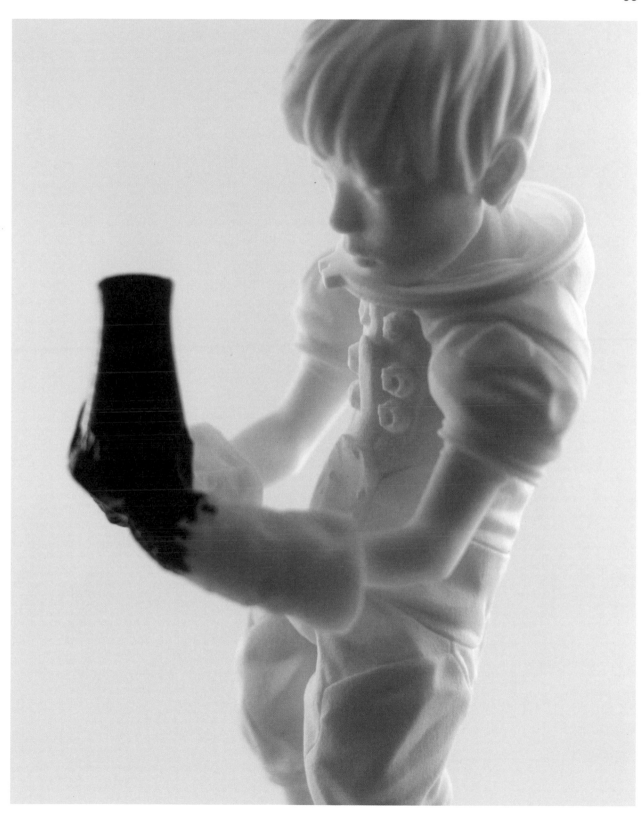

Fab Labs and the challenges ahead

62• CBA, 2012.

63• Anon, FabLabs.io. Available at: https://www.fablabs.io/ and Anon, Fab Lab Wiki - by NMÍ Kvikan. Available at: http://wiki.fablab.is/wiki/Main_Page

64• Anon, Fab Foundation. Available at: http://www.fabfoundation.org/ and Anon, Home - International Fab Lab Association. Available at: http://www.fablabinternational.org/

65• As many of these initiatives were short-lived, it is unpractical to list them.

66• Cutcher-Gershenfeld, J. (2007). Lateral Alignment in Innovation Networks. Presentation to FAB 4, the Fourth International fab lab Forum and Symposium on Digital Fabrication, Chicago, 19-24 August 2007. Available at: http://cba.mit.edu/events/07.08.fab/Cutcher-Gershenfeld.ppt

Despite the prime spot the term 'network' holds in the Fab Charter – it starts with the sentence 'Fab Labs are a global network of local labs'[62] – and the important functions the network is supposed to provide – 'operational, educational, technical, financial, and logistical assistance' – this network has still to develop. There are a few services the network offers to Fab Labs, principally a short directory listing global Fab Labs.[63] There are also a number of websites offering guidance for setting up Fab Labs[64], and a plethora of other sites aiming to promote exchange, to create business opportunities and to attract funding[65]. It has been acknowledged in the Fab Lab network that multiple forms of alignment – lateral, bottom-up and layered instead of top-down – are required, and that the network needs distributed leadership that is based on influence, not authority[66]. Yet many of the initiatives to strengthen the network are in actual fact authoritative approaches, above all when they try to become the central resource for a certain purpose, or to define what a Fab Lab is, once and for all – and this book is no exception.

The annual ritual of the network of gathering for an international Fab forum and symposium (or 'conference and festival' as it was called in Barcelona in 2014) is one established structure connecting the Fab Lab network. Growing attendance conceals the fact that these events risk losing out on broad, inclusive participation from the whole network. The price for attending is high when it involves international travel to far away countries – and for a large section of the Fab Lab population any destination is by definition far away. Spending ten days away is a substantial demand on the time budget of many Fab Labs. Remote participation at the Fab forum is virtually impossible, and while selected content might be available as a video stream, bandwidth at the receiving end might not be sufficient. It is a huge challenge for the Fab Lab network to become and remain inclusive, and not to hive off a Fab elite, and developing the sharing capabilities of the network is a burden that is still principally borne by the wealthier labs in the network.

67• Gershenfeld, 2006.

68• Stalder, F. 2013. The War of Data against Communication. Embros. Available at: http://embrostheater.blogspot.gr/2013/09/konrad-becker-felix-stalder-statistics.html

Another challenge Fab Labs and their networks face is their position regarding the social and political questions mentioned above. The louder voices in the Maker movement appear to side with the ideals of liberal individualism, projecting Makers as a new breed of Randian heroes. Is this picture of creative individualists who persevere with their goals – even when their ability and independence leads to conflict with others – really the vision Fab Labs want to pursue? As Fab Labs empower people through technology, they have to acknowledge that technology is a site of power. Consequently, the question needs to be asked: in whose name? If the Maker movement is indeed the final phase of winning the digital revolution[67], the earlier developments in this digital revolution are telling. The first decade of the Internet revolution (ca. 1995–2005) brought horizontality, cooperation and decentralization, as well as a vaguely anarchistic outlook. The second decade of Web 2.0, with its focus on data that put central control into the hands of unregulated corporations, is '[p]olitically speaking ... a counter revolution'[68].

This requires the development of a critical discourse around a few implicit assumptions: technology is not neutral but 'society made durable'[69], technology and people are 'entities that do things'[70], and technology comes with built-in societal, cultural and political assumptions; participation will not just work out-of-the-box, but is influenced by local cultural and social variables, such as heterogeneity and the role of elites; downward accountability and upward commitment are key to making participation work;[71] as Fab Labs and their network are at the forefront of technical innovation in and for society, they are also looked at in moral controversies to provide leadership, and not a 'neutral' hands-off attitude. Overall, in Morozov's analysis, 'there's more politicking – and politics – to be done here than enthusiasts... are willing to acknowledge'[72]. A particularly difficult case in point is the issue of funding of Fab Labs and their activities by large business corporations.

While still growing at an exponential rate, Fab Labs and their networks have to develop their practices of interaction and exchange. They have to replace top-down, center-out as the one single possible imaginable approach for organizing and experimenting, in favor of polycentric, bottom-up and lateral schemes. This in fact means that Fab Labs need to engage in constructing their practice and becoming institutions in 'a dialectic synthesis of what is going on in a society and what people are doing'[73]. They will need to avoid the potential enticement of corporate privatization and fab-washing. While being earnest – as an infrastructure for learning skills, developing inventions, creating businesses, and producing personalized products, and as a movement that is building its identity in a complex socio-technical and political-economical environment – Fab Labs must not forget that play is a crucial ingredient, nor forsake their non-utilitarian social role as third places, distinct from the first and second places of home and work[74], providing for civil society, democracy and civic engagement.

In the long term, Fab Labs have to prepare for a time when the concept has lost its novelty, when fabbing is not fabulous anymore. Depending on the decisions Fab Labs make about their purposes now and the routes they take in the near future, this could mean retiring to the position of consumer-oriented, commodity-producing facilities, or becoming part of a much broader development of the cultural, scientific and political configurations of society.

69• Latour, B., 1990. Technology is society made durable. The Sociological Review, 38(S1), pp.103–131.

70• Latour, B., 1994. Where are the Missing Masses? The Sociology of a Few Mundane Artifacts. In W. E. Bijker & J. Law, eds. Shaping Technology / Building Society: Studies in Sociotechnical Change. Cambridge, Mass.: The MIT Press.

71• Mansuri and Rao, 2004.

72• Morozov, E., 2014. Making It. Pick up a spot welder and join the revolution. New Yorker, January 13, 2014. Available at: http://www.newyorker.com/magazine/2014/01/13/making-it-2?currentPage=all

73• Sztompka, P., 1991. Society in Action: The Theory of Social Becoming. 1 edition., Chicago: University of Chicago Press.

74• Oldenburg, R., 1989. The Great Good Place: Cafes, Coffee Shops, Community Centers, Beauty Parlors, General Stores, Bars, Hangouts and How They Get You Through the Day. 1st edition., New York: Paragon House.

Massimo Menichinelli

The Business Dimension

How Fab Labs are created, maintained and financially supported

The concept of the Fab Lab as a space that provides citizens and communities with access to digital fabrication technologies could be seen as a social innovation. But such initiatives are not only social: they necessarily have business dimensions, even if some of them may be non-profit initiatives. At the same time, digital fabrication technologies are becoming not only more accurate and faster, but also cheaper and more interesting in a business sense. This chapter focuses on the different aspects of the business side of Fab Labs and digital fabrication, from the different types of Fab Labs, to their different budgets; from the process of developing a Fab Lab, to the main business models for projects developed inside Fab Labs. The profiles of three different Fab Labs explain further how different labs might have different business models.

Introduction

Fab Labs have emerged as the main forum for the democratization of digital fabrication: in them, people learn how to design and create, using tools and machines that bring together information technologies and physical processes and materials. In the Fab Lab network, bits and atoms and their interactions become the basis for empowered local communities, enabling them to form part of global networks of collaboration and sharing. Fab Labs started as a way of democratizing digital manufacturing technologies; they have expanded in many more directions, but the social dimension of Fab Labs is still of strategic importance. Fab Labs are therefore a very good example of social innovation, of an innovation that is ultimately social, but has its roots in technology – new scenarios emerge from public access to technology. Fab Labs are also an excellent example of a highly sought-after feature, namely, the ability to develop a social innovation that works locally, but that can at the same time be scaled up globally, without compromising its social content.

This fact highlights that Fab Labs are entities that are designed, developed and managed in different dimensions, and the business dimension is always present. Whether access to a Fab Lab is free, and regardless of whether it works on one specific type of project or more generally, it is ultimately a question of a business model and a business plan. Democratization and community-building need resources that have to be found, allocated and replenished continuously: the business aspect is always crucial. Business models and business plans are needed for Fab Labs, both for their institutional projects and for users' projects. Focusing on the business side of Fab Labs is also important if they are to have a real impact on specific localities and communities. The business side is probably not the most immediately attractive part (there are more profitable businesses than Fab Labs at the moment) but it is crucial for enabling social innovation. For example, having volunteers in a lab is very important for its ongoing development and success, but only a few small labs can rely exclusively on volunteering work.

More than ten years have passed since the first Fab Lab was founded, and even if this may seem like a long time, the development of the Fab Lab network has grown almost exponentially, and this boom is a quite recent phenomenon. While in the first period Fab Labs tended to be similar, they have of late become increasingly differentiated and specialized. We therefore have more data about some formats and business models, and less about others: we know much more about how to design and manage a lab hosted by an institution, for example, and much less about independent labs.

1959 CNC machine: Milwaukee-Matic-II was the first machine with a tool changer operated by punched tapes computer.

Moreover, both digital fabrication and the Maker / DIY markets have been emerging recently: there are many optimistic predictions about present and future growth, but we still have to understand how large and relevant these markets will be, which sectors will be impacted by them, and which countries and regions will play a leading role in them. The good news is that there has never been a better moment to take an active role in this exploration, and to design or discover future scenarios for both society and the economy.

The business side of Digital Fabrication

Digital Fabrication technologies have been developed since the 1950s, when the first manufacturing machines (milling machines) were connected to computers that could control them numerically (hence the term CNC: computer numerically controlled). Since then, the same concept of moving a manufacturing tool with a numerical control has been applied to many different tools, processes and technologies, giving rise to a complete ecosystem of manufacturing tools. Fab Labs were created in order to democratize access to the cheapest and easiest to use of these technologies. Some of them, like CNC milling, are therefore quite old, and are now a stable set of technologies manufactured by established companies, while other technologies, such as 3D printing, are emerging with a fast-paced evolution of rising quality and falling-off of prices. This is why there is currently a lot of attention being paid to 3D printing as the future of manufacturing (and not just prototyping), and why the business of 3D printing is growing as fast as it is. Each year a Wohlers Associates report tracks the development of the 3D printing industry. In the latest edition of the report[1] they showed that the market for 3D printing grew 28.6 % in 2012 to $2.204 billion, up from $1.714 billion in 2011 (29.4 % growth). The average annual growth of the industry over the past 25 years is at 25.4 %. Not surprisingly, the growth of the more recent low-cost (under $5,000) 'personal' or 'desktop' (or 'democratic', if you prefer) 3D printer market was on average 346 % each year from 2008 through 2011. By 2017, Wohlers Associates believes that the sale of 3D printing products and services will approach $6 billion worldwide, and $10.8 billion by 2021.

1• Wohlers, T.T. & Wohlers Associates, 2013. Wohlers report 2013: additive manufacturing and 3D printing state of the industry: annual worldwide progress report.

While Fab Labs tend to rely on these 'personal', 'desktop' or 'democratic' technologies, there are also commercial services that 'democratize' (even if a fee is involved) access to most advanced 3D printing technologies. It is now possible to upload, create or customize 3D models on online platforms, which then 3D print your model in steel, silver, brass, gold, glass, ceramic, full color plastic, rubber, and more. These platforms can send the physical object to you, or sell it online for you by creating an e-commerce store for your products: you can now access a factory, and manufacture and sell your own products simply by paying them a percentage. Platforms like Shapeways[2] (The Netherlands and USA), Sculpteo[3] (France) or i.materialise[4] (Belgium) constitute a representative example of the cloud manufacturing ecosystem of 3D printing, by way of which anybody can access these technologies through a web interface, and where the price of a product depends on the actual volume of material used, plus shipping. But cloud manufacturing is not only about 3D printing: companies like Ponoko[5] (New Zealand) offer not only 3D printing but also laser cutting (for a while it also offered CNC milling and the embedding of electronic components, thanks to a partnership with Sparkfun[6]). Ponoko is not just a central factory, but is also a distributed network: five digital factories have been established since 2008 in Wellington, San Francisco, Berlin, Milan and London. Each hub is locally owned and operated; pricing, materials, support, delivery and business terms may thus vary between them. And there are even more distributed platforms that link all the users that have a CNC milling machine (100k Garages[7]) or 3D printers (3D Hubs[8]) and that are available for fabricating digital files uploaded by users, who usually choose the closest facility for fabrication.

There are several ways to build networks of digital fabrication technologies, using a scenario that we could call Distributed Manufacturing: by establishing labs open to the general public (Fab Labs); by establishing cloud manufacturing as factories with a web interface (Ponoko, Shapeways, Sculpteo, i.materialise); and by networking already distributed machines and resources (100k Garages, 3D Hubs). All of these strategies are implementing the first steps in bringing about a new Industrial Revolution in manufacturing. Their economic potential will bloom in the near future, but we can already observe some of its manifestations, for example by considering the history of Shapeways. Founded within the Philips Lifestyle Incubator at the beginning of 2009, Shapeways then became independent, and in 2010 it received $5,000,000 from VC Index Ventures and Union Square Ventures (the same firm that backed Twitter) to open a manufacturing facility and headquarters in the USA[9]. After that it received another $30,000,000 in 2013[10], showing the growing importance of, and interest in, distributed manufacturing.

While the distribution of such digital fabrication technologies will represent a possible new Industrial Revolution through Distributed Manufacturing, their impact can best be understood if we take into account the integrated ecosystem of different technologies and

2• Shapeways, Inc., Shapeways - 3D Printing Service and Marketplace. Available at: http://www.shapeways.com/

3• Sculpteo, Sculpteo | Your 3D design turns into reality with 3D Printing. Available at: http://www.sculpteo.com/en/

4• i.materialise, 3D Printing Service i.materialise | Home. Available at: http://i.materialise.com/

5• Ponoko Limited, Laser cutting and engraving – design, make & build your own products with Ponoko. Available at: https://www.ponoko.com/

6• SparkFun Electronics, SparkFun Electronics. Available at: https://www.sparkfun.com/

7• ShopBot Tools, Inc., 2012. 100kGarages - Where projects are made by digital fabricators (fabbers) working with 2-D or 3-D digital fabrication tools. Available at: http://www.100kgarages.com/

8• 3D Hubs B.V., 3D Hubs: Local 3D printing services and 3D Printers. Available at: http://www.3dhubs.com/

9• Menichinelli, M., 2011. openp2pdesign.org. Business Models for Fab Labs. Available at: http://www.openp2pdesign.org/2011/fabbing/business-models-for-fab-labs/

10• Copeland, M.V., 2013. The $30M Bet That Shapeways Becomes a Factory for Everyone | Business. WIRED. Available at: http://www.wired.com/2013/04/30-million-bet-that-shapeways-becomes-a-factory-for-everyone/

3D printed bow tie.
Image courtesy of Shapeways

businesses. Not all technologies and businesses will have a profound impact on society and markets, but their integration undoubtedly will. There are still differences between what can be achieved with a Fab Lab and with a bigger factory, with cloud manufacturing or desktop manufacturing: a Fab Lab could replace a factory, but only if it receives a similar level of investment; desktop manufacturing will not be able to compete with cloud manufacturing factories, unless it adopts the same technologies and facilities. But the integration of all these aspects will provide different solutions to different needs, from the repair of a single object to the gadget, from a small series of objects to a distributed mass scale series of products. And since research on digital fabrication is advancing alongside nanotechnology and biotechnology[11], there are even more extensive possibilities and directions for future businesses related to digital fabrication and ICT, from the sourcing of materials to the design, manufacturing and distribution of end products.

11• Menichinelli, M., 2013. The bio future of Design and Fabbing. openp2pdesign.org. Available at: http://www.openp2pdesign. org/2013/fabbing/the-bio-future-of-design-and-fabbing/

Starting a Fab Lab: developing a project and its business

Fab Labs are therefore one of the key components of the Distributed Manufacturing scenario, where multiple digital fabrication technologies are distributed, yet networked and coordinated only when necessary. More specifically, they are not just technologies: they are places for the gathering together of local communities, much more so than desktop manufacturing, cloud manufacturing or traditional factories. Fab Labs should then be understood not just as a set of technologies, but as places for communities with a specific size and local dimension. The first step to consider before thinking about the business of a Fab Lab is thus to choose the specific size that we are going to work with. Leading through inspiration, on many occasions Neil Gershenfeld has proposed a framework for understanding the possible structure and size of Fab Labs. This concept could be helpful in understanding Fab Labs, and also in orientating their growth and diffusion. Adopting a 'power of 10' scale approach (inspired by a famous video made by Charles and Ray Eames[12]), Gershenfeld has proposed that Fab Lab size should roughly follow a 'power of 10' scale of investment, with regard only to technologies, machines and materials. For example, he has been developing an inventory[13] that accounts for around $100,000 (also written as 100K) of technologies, components and materials. This inventory is of course not compulsory for other Fab Labs, but serves more as an example and recommendation of how a typical Fab Lab should operate: machines and technologies are chosen on the basis of their performance over price ratio, their supply chain and worldwide availability, and on the relationships of the vendor with the Fab Foundation and the Fab Lab network. During FAB9, the international event for Fab Labs that took place in Yokohama, Japan, in August 2013, the work of Bart Bakker on Mini Fab Labs became part of the Fab Lab network, with a specification of the $/€10,000 (10K) Fab Lab inventory[14]. No specifications have yet been proposed for $/€1000 (1K) Fab Labs or $/€1,000,000 (1000K) Fab Labs: there will, however, be more opportunities for discussing these in the future, as more compact and multi-purpose technologies become available for 1K Fab Labs, and as more opportunities for bigger labs with a wider business and community reach will arise for 1000K Fab Labs. At the moment, the 1K Micro Fab Lab falls into the desktop manufacturing category, the 10K Mini Fab Lab falls into the block/association category, the 100K Medium Fab Lab falls into the neighborhood/institution category, and the 1000K Mega Fab Lab falls into the city/TechShop (a franchising of much bigger makerspaces) category. These categories are useful for coordinating the network and enabling more labs, even when there are fewer resources available, or the context requires and permits a bigger lab. But we always have to remember that the scale is approximate, and that it should be regarded more as a reference framework than as an exact prescription; machines and

12• Eames, C. & Eames, R., 1977. Powers of Ten, Available at: http://www.youtube.com/watch?v=0fKBhvDjuy0.

13• CBA, fab lab inventory. Available at: http://fab.cba.mit.edu/about/fab/inv.html

14• The discussion regarding the 10K Fab Lab is hosted in this Facebook group: Anon, Small Fab Lab Network. Available at: https://www.facebook.com/groups/smallFabLab/

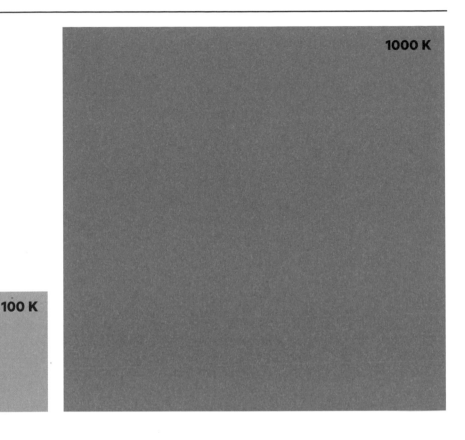

1K *MICRO FAB LAB*
DESKTOP MANUFACTURING

↗ **10K** *MINI FAB LAB*
BLOCK/ASSOCIATION

↗ **100K** *MEDIUM FAB LAB*
NEIGHBORHOOD/INSTITUTION

↗ **1000K** *MEGA FAB LABS*
CITY/TECHSHOP

technologies might have a different price in each country, while a different preference for a specific machine or technology could change the balance. The inventories take into account only technologies and materials, and not other relevant expenses such as spaces and wages. If we take into account wages, space costs, service costs and other expenses, the amount approximately doubles, so that 10K becomes 20K, 100K becomes 200K and 1000K becomes 2000K (I've noticed this in my practice as a Fab Lab developer and researcher, and TechShops have similar levels of investments[15] regarding 1000K). As mentioned above, desktop manufacturing cannot compete with Fab Labs on the scale of production (but can be more distributed), so the different levels in the power of ten of Fab Labs cannot be compared. A 1K Micro Fab Lab is on the same scale as desktop manufacturing, 10K Mini Fab Labs are an intermediate level, while 100K Medium Fab Lab is a good level for small community and personal fabrication, and 1000K Mega Fab Lab is the right size for a substantial amount of manufacturing, roughly equivalent to the production of a factory of the same size. But this framework still gives us no more than a reference tool for understanding the scale of investment and the nature of the labs that we would like to develop.

15• Hurst, N., 2014. TechShop's Not-So-Secret Ingredient. MAKE. Available at: http://makezine.com/magazine/make-40/techshops-not-so-secret-ingredient/

Now that we know the scale of investment and the structure of a possible Fab Lab, we need to define its project. There are many different dimensions to consider, and the business model and the business plan are the documents that cover the financial dimension. A business model describes the rationale of how an organization creates, delivers, and captures value – how will it make money and sustain its profits over time? A business model therefore documents the strategy of an organization, its goals and the direction it wants to take. In addition, the business plan sets out (and calculates) which paths and steps the organization should take to achieve its goals. It is the blueprint for the decision-making behind the implementation of the strategy laid out in the business model. This is an important distinction to understand: both documents need to be designed and balanced at the same time. We cannot start calculating the details of the business plan before choosing the business model. For Fab Labs, a typical business model falls between 'for profit' and 'not for profit' models: there are different possible proportions that can be implemented, and each Fab Lab has its own mix. For this reason, the difficult part of drafting the business model and harmonizing it with the business plan is finding the right balance between the financial side and the social side. As a consequence, a Fab Lab's project is not fully developed with just the business model and business plan: in my experience as a Fab Lab developer (three labs I've developed are covered in this chapter), it is critical to understand the potential users, their activities within the lab, and the local context. After having identified the potential users, it is possible to design the services of the lab, and ultimately define technologies, tools and machines for it. Ideally, the identification of the users and the design of the activities, services and technologies should be carried out at the same time as the design of the business model and business plan, with feedback from each of these multiple documents and design activities.

Since the Fab Lab format is an interesting and successful example of an open source organization, there is some flexibility built into the format itself, and each Fab Lab is a bit different from the others. As a consequence, each lab has its own business model. There are, however, common patterns. For example, there are four conditions[16] that a lab needs to fulfill in order to be called a Fab Lab.

These aspects show that Fab Labs are not directly defined by a specific business model. There are many possible combinations of revenues and expenses, and therefore different business models, and the Fab Charter only specifies (at the moment) that businesses developed in a Fab Lab should grow beyond rather than within the lab. In recent years, common business models and patterns of elements for business models have been identified. For example, four business models for Fab Labs have been discussed on the FabWiki[17]:

16• Fab Foundation, What qualifies as a Fab Lab? Available at: http://www.fabfoundation.org/fab-labs/%20fab-lab-criteria/

17• Troxler, P., 2010. Proposal - Fab Lab Wiki - by NMÍ Kvikan. Available at: http://wiki.fablab.is/wiki/Proposal

Necessary Conditions for a Fab Lab

There should be public access to the lab.	The lab should share a common set of processes, tools and technologies with the other labs.
The lab should subscribe to and follow the Fab Lab manifesto, the Fab Charter.[18]	The lab should participate in the global network of Fab Labs. It cannot be isolated from the network.

18• CBA, 2012. The Fab Charter. Center for Bits and Atoms. Available at: http://fab.cba. mit.edu/about/charter/

Table 1: Criteria for the qualification of Fab Labs (Source: Fab Foundation)

The Four Main Business Models of Fab Labs

The Enabler business model

launch new labs or provide maintenance, supply chain or related tools and services for existing labs, in order to strengthen the network;

The Education business model

a globally distributed model of education through Fab Labs (with the Fab Academy), where global experts in particular topics can deliver training from local Fab Labs or even from universities/businesses via the Fab Lab video conferencing network. Peer-to-peer learning among users is also part of this business model;

The Incubator business model

provide infrastructure for entrepreneurs to turn their Fab Lab creations into sustainable businesses. The incubator provides back-office infrastructure, promotion & marketing, seed capital, leverage in the Fab Lab network, and other venture infrastructure to enable the entrepreneur to focus on her areas of expertise;

The Replicated / Network business model

provide a product, service or curriculum that operates by utilizing the infrastructure, staff and expertise of a local Fab Lab. Such opportunities can be replicated, sold and executed at many (or all) local labs, with sustainable revenue at each location. The leverage of all labs in the network simultaneously promoting and delivering the business creates strength and reach for the brand.

Table 2: Four business models for Fab Labs (Source: Troxler, 2010)

19• Troxler, P. & Wolf, P., 2010. Bending the Rules: The Fab Lab Innovation Ecology. In CINet Conference 2010. Zurich. Available at: http://www.continuous-innovation.net/Events/CINet2010/downloads.html

In 2010, Peter Troxler and Patricia Wolf proposed another analysis of possible business models for Fab Labs[19]. They identified four possible models (Table 2), based around the intersections of open and closed intellectual property, and Fab Lab as facility or as innovation support. This is one of the many possible segmentations of digital fabrication or making labs: there is not always a clear cut distinction between closed and open intellectual property, since in many labs the two models often overlap. It is very useful, however, for understanding the typical differences between Fab Labs (open intellectual property, the lab as a facility or for the development of innovations) and other making workshops (closed intellectual property and lab as a facility) or design / engineering firms (closed intellectual property and lab for the development of innovations). There may still be overlaps: some Fab Labs could also act as design / engineering firms besides being open to the general public.

Business Models for Fab Labs

	LAB AS FACILITY	→ INNOVATION LAB
CLOSED IP ↓	Traditional machine shop	Typical innovation consultancy
OPEN IP	Typical Fab Lab approach	Fab Lab innovation ecology

Table 3: Four business models for Fab Labs (Source: Troxler, Wolf, 2010)

These business models can be built around the eight business patterns[20] that have been identified so far on the FabWiki:

20• Troxler, P., 2012. Business Patterns – Fab Lab Wiki – by NMĺ Kvikan. Available at: http://wiki.fablab.is/wiki/Business_Patterns

Business Patterns for Fab Labs

Grant-based funding provided by a public or private organization	Embedded in (educational) institutions the lab is hosted by a public or private organization, most often an educational one	Co-X (co-working, sharing infrastructure) renting idle infrastructure and equipment	Operating as a prototype shop building prototypes (or final products) as a service
Access fees charging a fee for the access to the lab	Educational activities workshops and courses led by local and global instructors	Techno tourism event- or activity-based fees	Gurus for hire consultancy or support fees

Table 4: Business patterns for Fab Labs (Source: Troxler, 2012)

These different business models and patterns show that Fab Labs can adopt different directions: from being hosted by, or funded by a grant from an organization, to being completely independent; from facilitating other people's projects, to designing and manufacturing their own; from distributed and peer-to-peer education, to distributed and open source manufacturing. These are all the different revenues that can be balanced in a business plan. It is critical to take into account that the more relevant (financially) a source of income, the more important users that represent that source will be. For example, if the lab is hosted by a university, the students and the staff will be the most important user base; if the lab is independent, paying or associate users will be the most important user base; and so on. The importance of a lab's user base brings us to another consideration regarding the business model and the business plan: even if they are based on data and well-thought-out design, they still constitute a hypothesis that needs to be proved and/or refined. Therefore, the first 6–12 months of a lab, after its inauguration and opening to the public, are vital for discovering how much of the business model and plan are accurate, who the actual users are, and what their needs are. The business model and business plan should be thought of, then, as an integrated document, and its balance should be refined in the first year of the lab's life. Furthermore, the business model and business plan should consider all the hidden costs of launching and running a lab, including, for example, wages, rent and expenses in terms of space, services, consulting, materials, components and machines.

Profile 1 ▪ Aalto Fablab

HELSINKI − FINLAND

HTTP://FABLAB.AALTO.FI/SITE/
HTTP://MEDIAFACTORY.AALTO.FI/
HTTP://WWW.AALTO.FI/EN/

The Aalto Fablab (Helsinki, Finland) is part of Aalto University. Established in 2010 in Helsinki and Espoo (Finland), Aalto University was born out of a merger of the Helsinki School of Economics (established in 1904), Helsinki University of Technology (founded in 1849) and the University of Art and Design, Helsinki (founded in 1871). In order to facilitate the transition from three different universities to one university, four different Factories were established: Design Factory, Media Factory, Service Factory, Health Factory. These factories provide platforms for collaboration and development outside the usual scope of academic departments and research units. They enable collaboration among different units, different faculties and external organizations, such as companies or public / third sector institutions.

Aalto Fablab was developed in Spring 2011 and inaugurated in June 2012 as part of Aalto Media Factory, a decision which highlights the importance of digital fabrication as a set of new media that can foster collaboration among different stakeholders. The Aalto Media Factory focuses on developing multidisciplinary, media-related research and education, by supporting joint ventures, such as research projects, course pilots and event productions, and by providing funding, coaching, tools and spaces. The Aalto Fablab and its related Electronics Studio can use the infrastructure of Aalto Media Factory, which includes a kitchen, meeting rooms, auditorium, video editing room and a web development support service. Three people work at Aalto Fablab, each with a different background and role. Users are normally students and staff of the University (especially from the art, design and media departments), but at least one day a week the lab is open to anyone. Being part of a Finnish university, all the expenses of the lab are covered by the Aalto Media Factory and therefore the Aalto University. The Finnish welfare state is extremely generous (which in itself represents another variable to be considered): using the machines is free of charge, as is hands-on assistance, although any use of materials above a limit of €5 must be paid for by the user.

Profile 2 ▪ MUSE FabLab   **TRENTO – ITALY**

HTTP://FABLAB.MUSE.IT/   HTTPS://SEESCIENCE.EU/
HTTP://WWW.MUSE.IT/EN/PAGES/DEFAULT.ASPX
HTTP://WWW.FONDAZIONECASSARURALEDITRENTO.IT/FONDAZIONE.ASPX

The MUSE FabLab (Trento, Italy) is part of the MUSE Science Museum, an auxiliary body of the Autonomous Province of Trento. The mission of the museum is to interpret nature, starting with the mountains, using the tools and applications of scientific research, thereby stimulating scientific curiosity and the pleasure of learning, as well as highlighting the value of science, innovation, and sustainability. The roots of the Science Museum can be found in several private collections from the late 1700s, which were merged to establish the museum in 1922. The growth of the audience and the importance of the museum led to the development of a new building and a new organization; the Tridentine Museum of Natural Sciences became the MUSE Science Museum. The new museum project started in 2002 and was approved by the provincial government in 2006. The project was transformed into an architectural plan and entrusted to Renzo Piano, who designed the building and acted as artistic director for the stands. On the 27th of July 2013, the new MUSE Science Museum (and the MUSE FabLab) opened to the public in its new headquarters, located in the quartier Le Albere, an urban redevelopment project that was also designed by Renzo Piano.

The MUSE FabLab has a notable funding pattern, in that all of the resources that enabled the lab to start came from three organizations: the MUSE Science Museum, Fondazione Cassa Rurale di Trento (the foundation of a local bank that works on local community development) and the SEE Science project (a South East European project that networks Science Centers in order to increase public awareness of the importance of natural sciences, technology and innovation as key determinants of economic growth). The aim of the lab is to provide the tools and experience for helping users acquire understanding of both science and the quest for a sustainable society. The budget for developing and launching the lab, then, was provided by these organizations; the future budget will be based on users, visitors to the museum and partnerships with research centers and local companies. Two people run the lab, and the users are mainly students and families that come to visit the museum; other users such as local researchers, entrepreneurs and Makers have been considered.

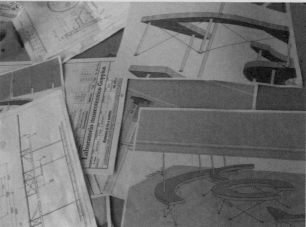

Profile 3 ▪ opendot

MILAN — ITALY

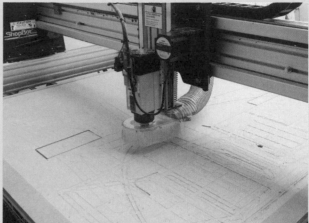

Opendot is a lab launched by dotdotdot in Milan, Italy. Founded in 2004, dotdotdot is a multidisciplinary design studio that combines art, architecture, construction and design, infusing them with new technologies and new media. Most of the projects developed by dotdotdot are interaction design or exhibition design projects, both types being based on new technologies, electronics and digital fabrication. dotdotdot has also participated actively in open source communities for years. It was natural, then, that they should have decided to launch a Fab Lab: it is independent of the studio, and based on a commercial company and a cultural association, one handling the 'for profit' and the other the 'not for profit' sides of the lab. The lab is run by two people. Its business model most resembles that of coworking (users need to be registered in order to access the lab), and the main users are design companies (which are offered custom workshops for the development of new and open source projects), professionals (graphic design, product design, fashion design, interaction design, photography and movie design), and also students and retired Makers. The lab was launched in September 2014: this is an example of a private and independent lab which was founded by a pool of private associates.

Business models for Fab Lab projects

A very common question regarding Fab Labs is how to craft the right business model to make them financially sustainable. There is not, however, a single easy answer: it is a matter of creating a project that is tailored to the local context, while at the same time being connected to the global network. Fab Labs are not a franchise in which everything is codified into a manual: local adaptability is what is making them successful and enabling them to spread everywhere. Furthermore, Fab Labs are not an established business format with a long tradition; there is still a lot of experimentation to be done, both at the level of designing and opening up a lab, and at the level of the user, regarding how to use and influence the lab. This same flexibility applies to the projects developed in a Fab Lab: there are many different possible directions, depending both on the inventor and on the local nature of the lab.

For example, regarding the intellectual property of projects developed in a Fab Lab, the Fab Charter allows great flexibility, since '[D]esigns and processes developed in fab labs can be protected and sold however an inventor chooses, but should remain available for individuals to use and learn from'[21]. This means that projects developed in Fab Labs are often but not always developed and published as open source (open source software, open hardware, open design). Sometimes users share only the documentation of the processes needed for designing and manufacturing the projects, while sometimes the projects adopt a traditional closed intellectual property strategy. When the project is developed and shared as open source, this can be with both non- and for profit models. On the one hand, this might be solely for ethical issues, for teaching, or to acquire a reputation among the community or with companies. On the other, there could be monetary transactions involved in selling blueprints, manufactured objects or custom versions, or for more specific services developed around the projects[22]. Generally, an open source business model implies that intellectual property must be shared, or even if it is sold, anybody should still be able to access and improve or share it. Otherwise, there are business models that are based on monetary transactions that regulate the ownership of intellectual property, yet exchange in a way that is more open relative to traditional business models. These models are normally referred to as Open Innovation: we still have more open boundaries in organizations, but here intellectual property is a good to be exchanged with a monetary compensation, and not shared publicly[23]. This is still a possibility for the projects developed in Fab Labs, especially when there is an interest in them from traditional companies that are not yet ready for open source business models. Likewise, projects could also be co-designed by users, companies and professional designers, in the expanding field of co-design and co-creation[24]. And let's not forget that

21• CBA, 2012.

22• For a good explanation of Open Source business and organization models, see Weber, S., 2005. The Success of Open Source, Harvard University Press. For a good introduction to open design, see Abel, B. et al., 2011. Open Design Now: Why Design Cannot Remain Exclusive, Amsterdam: BIS Publishers. Available at: http://opendesignnow.org/

23• You can read more on this topic in Chesbrough, H.W., 2003. Open Innovation: The New Imperative for Creating and Profiting from Technology, Harvard Business School Press and Chesbrough, H., 2011. Open Services Innovation: Rethinking Your Business to Grow and Compete in a New Era 1st ed., Jossey-Bass.

many designers, architects and engineers are users of Fab Labs: it is therefore difficult to tell when a Fab Lab is used by a professional or not. From this fact stems the wide range of possibilities we have covered, from the more traditional to the more open business models.

As we have seen, there may be many different ways for users of the Fab Labs to develop projects for themselves or for/with a company. What it is important to understand about Fab Labs is that by democratizing access to tools, and by gathering together local and collaborative communities, many more projects and solutions can be arrived at than by working only with companies and professionals. We are, essentially, multiplying opportunities for innovation. There has been a lot of research on innovation developed by users and not companies in the past decades, especially the work of Eric von Hippel[25]. For example, he found that 6.2% of UK consumers (that is, around 3 million individuals) engage in consumer product innovation. In order for traditional companies to acquire the same amount of distributed innovation, they would have to spend 2.3 times what they already spend on R&D. These studies have proved that users, individually or in communities, develop a relevant amount of projects and innovation, and Fab Labs could help them work and collaborate among themselves and with companies. Fab Labs could then be both a place for citizens' personal manufacturing, but also a place where professionals and citizens work together on common projects, to be manufactured in Fab Labs or in local companies. Such places could then become a meeting point not just for a few Makers, but for citizens, professionals, researchers, educators and companies, creating a local hub for education, research and innovation, especially for small institutions like local schools or small and medium enterprises. In this way, Fab Lab could have a critical role in improving localities and their communities, at both the social and financial level.

24• Sanders, E.B.-N. & Stappers, P.J., 2008. Co-creation and the new landscapes of design. CoDesign: International Journal of CoCreation in Design and the Arts, 4(1), p.5.

25• Von Hippel, E., 2005. Democratizing innovation, Cambridge, Mass.: MIT Press. Available at: http://web.mit.edu/evhippel/www/democ1.htm. Von Hippel, E., 1988. The sources of innovation, New York: Oxford University Press. Available at: http://web.mit.edu/evhippel/www/sources.htm. Hippel, E.A.V., Jong, J.D. & Flowers, S., 2010. Comparing Business and Household Sector Innovation in Consumer Products: Findings from a Representative Study in the UK. SSRN eLibrary. Available at: http://papers.ssrn.com/sol3/papers.cfm?abstract_id=1683503

Cecilia Raspanti
Alex Schaub

The Creative Process

A sample project from Fablab Amsterdam

Generally, Fab Labs have many kinds of machines, with the focus on covering the most important processes. The range of available processes is quite wide, and still expanding. Each user has her own expertise, approach and goals, which influence the projects she develops in the lab. In order to understand the possibilities of Fab Labs, in this chapter we follow the path of a designer and her experience in learning to work in a Fab Lab. While experimenting with laser cutters, big CNC wood routers and 3D printers she developed a full fashion collection that includes clothes, jewellery, bags and even custom weaving and knitting looms. This process gives us an idea of how fashion design can adopt digital fabrication, from the manufacturing of finished goods to the manufacturing of custom machines for more experimental processes and goods.

Typical design process in a Fab Lab

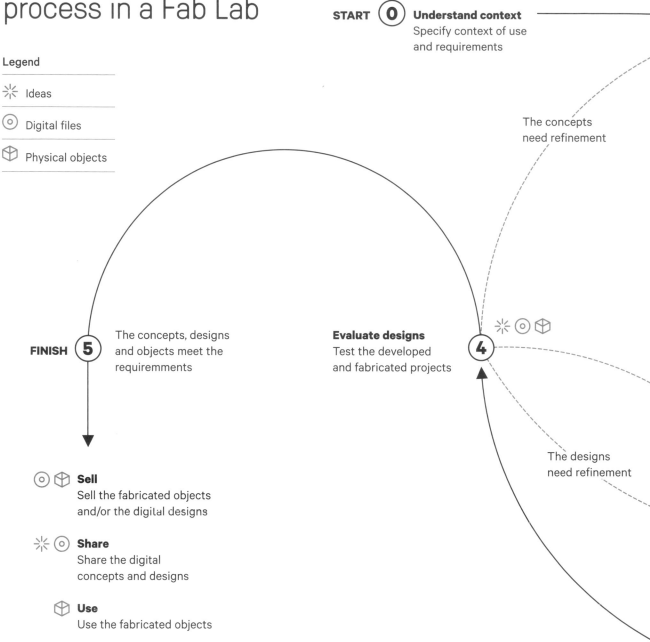

START (**0**) **Understand context**
Specify context of use
and requirements

Legend

☀ Ideas

◎ Digital files

⬡ Physical objects

The concepts
need refinement

FINISH (**5**) The concepts, designs
and objects meet the
requiremments

Evaluate designs
Test the developed
and fabricated projects

☀ ◎ ⬡

(**4**)

◎ ⬡ **Sell**
Sell the fabricated objects
and/or the digital designs

☀ ◎ **Share**
Share the digital
concepts and designs

The designs
need refinement

⬡ **Use**
Use the fabricated objects

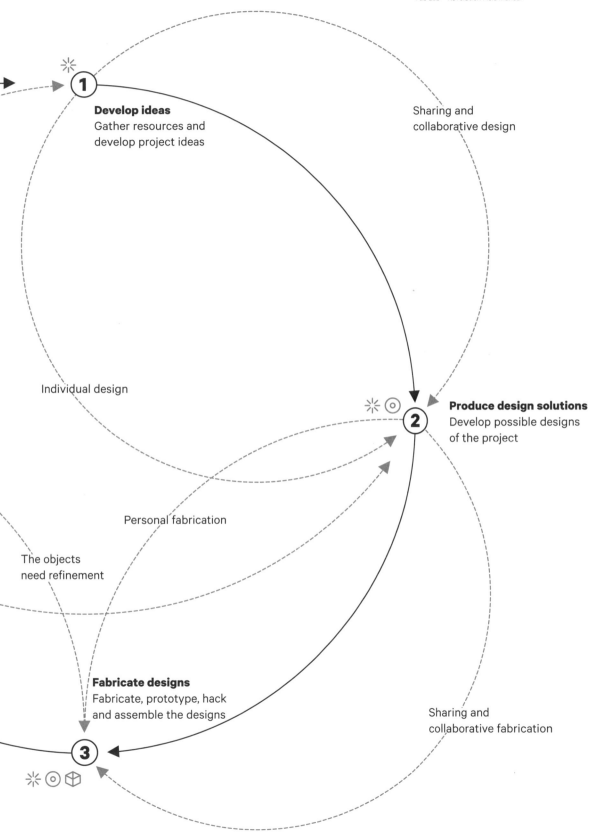

Develop ideas
Gather resources and
develop project ideas

Sharing and
collaborative design

Individual design

Produce design solutions
Develop possible designs
of the project

Personal fabrication

The objects
need refinement

Fabricate designs
Fabricate, prototype, hack
and assemble the designs

Sharing and
collaborative fabrication

Fablab Amsterdam – best practice by Alex Schaub

As director of the Fablab Amsterdam for more than six years, I always hugely enjoyed the open days. On open days, when everybody is welcome to create and experiment, people bring an incredible diversity of projects into Fab Labs, from architectural models to musical instruments, from fashion designs to electronics, and from product design to healthcare applications. The mix of all these people working simultaneously in our Fab Lab makes it a hugely vibrant place.

In February 2013, Cecilia Raspanti came to visit us, to do some CNC embroidery. I was impressed that I didn't have to explain to her how the process works: she simply figured it out herself, using our instruction video and other documentation on our website[1] – as well, of course, as her natural talent. While the embroidery machine did her work, Cecilia strolled around the lab and was very curious about the gear we have here. She kept reserving machines and soon became a regular visitor. Over the next few months, she explored all kinds of machines and processes and became even more inspired. I loved to see how she was growing into the Fab Lab world, and her fascination with it was written on her face. As I also noticed that she had talent and considerable knowledge of craftsmanship within the fashion segment, and wanted to combine it with new digital fabrication, I invited her to undertake a three month internship, so she could spend more time researching for her new world of digital fabrication, and learn the implications of CNC machining, among other processes.

What followed was an impressive line of a fashion collection, which took her about eight months to complete[2]. It was great to inspire her with new techniques and tricks, and to see her growing from a rookie to an expert, whether in the software domain or CNC machining.

I am delighted to share the discoveries regarding the materials and techniques she explored during these months; in my opinion, the final result is a masterpiece of beauty in craftsmanship and digital fabrication. I wouldn't be surprised if these dresses appear on the market.

1• Anon, Home | Fablab Amsterdam. Available at: http://fablab.waag.org/

2• Cecilia Raspanti's project is documented on Raspanti, C., Ceciilya. Available at: http://ceciilya.com/ and Raspanti, C., 2014. Cecilia's Internship project | Fablab Amsterdam. Available at: https://web.archive.org/web/20150827043417/http://fablab.waag.org/project/cecilias-internship-project

Opposite page:
Ernst Haeckel, Kunstformen der Natur, Tafel 9. Hexacoralla, Meandrina.

A research project in a Fab Lab by Cecilia Raspanti

Inspiration

Cover of the book Kunstformen der Natur, 1904 edition.

3• Haeckel, E., 1904. Kunstformen der Natur, Leipzig, Wien: Verlag des Bibliographischen Instituts. Available at: http://caliban. mpiz-koeln.mpg.de/~stueber/haeckel/kunstformen/natur.html

4• For more information regarding radiolarians, see Anon, 2014. Radiolaria. Wikipedia, the free encyclopedia. Available at: http://en.wikipedia.org/w/index.php?title=Radiolaria&oldid=635568782

Voronoi tesselation diagram.

This research project started with the aim of better understanding and exploring the possibilities of combining fashion and textile design with digital fabrication. Personally, as a fashion designer, the world of Fab Labs and digital fabrication was practically unknown to me until I started this project. I had previously worked mostly on textile manipulation from a craftsmanship point of view, researching old techniques and merging them together to create new combinations of materials to work with. Tools are what we need to make these new combinations possible, and the switch from purely mechanical machines to computer-guided ones opened infinite possibilities, and was a continuous inspiration in my discovery process.

Nature has always been an incredible source of inspiration; it changes and evolves, starts and finishes, in the same way as we ourselves and our ideas and projects do. Large numbers of artists, designers and creatives in general have been attracted to the combinations of soft and hard, fluid and rigid, that nature presents. My research started from the particular point of view on nature of an artist and scientist, Ernst Haeckel, and the book he published between 1899 and 1904, Kunstformen der Natur (Artforms in Nature)[3]. One of the most interesting aspects of his publication is that it's clearly addressed to a wider public, with its colorful and charming illustrations, rather than to specific groups of scientists and scholars. In his work, Haeckel documented all the species he encountered, discovered and studied over the years, especially focusing on his main obsession, marine invertebrates such as Radiolarians. Radiolarians are marine planktonic organisms widely distributed in all the oceans and present at almost all depths[4]. They are well-known for their elaborate siliceous skeleton, recognizable for its radial symmetry and its incredible variety of different perforated patterns. These mineral skeletons are formed through the absorption of silica, and their size ranges from 30 microns to almost 2 mm. Haeckel's aim was not only to understand these organisms and to study their structures and functional ways of evolving, but also to show the beauty and complexity in their almost infinite combination of shapes. That's why one of the aspects that characterizes the illustrations of Ernst Haeckel is the mathematical geometry and symmetry that we find in all of his work. Most of the aggregate structures he describes seem to be based on Voronoi patterns, which had been informally analyzed throughout the whole of the second half of the 19th century, while the Voronoi diagram itself, named after the mathematician Georgy Feodosievych Voronoy, was actually only defined around 1908. The incredible structural and visual properties of Voronoi were really interesting to analyze

Ernst Haeckel, Kunstformen der Natur,
Tafel 43. Nudibranchia, Aeolis.

Ernst Haeckel, Kunstformen der Natur,
Tafel 17. Siphonophorae, Porpema.

during the project, and resulted in a perfect match with the machine and computational work that define digital fabrication tools. Haeckel's research was very much a reflection of the two focii that characterized him as a scientist. On the one hand, he was a scientist, analyzing the systems that belong to nature and to the outer world he discovered through his research. On the other hand, however, he was an artist, attracted by beauty, poetry, feelings, and everything related to a more internal and personal world. Radiolarians were for him what brought these two often contradicting and conflicting worlds together. In this project, machines and digital fabrication represent that systematic, controlled world, and the craftsmanship aspect and artisanal handcrafted techniques represent the other. Understanding and exploring the possibilities of this contrasting combination, and also finding a more fluid relation between the two, became part of the project, as well as one of its aims. The incredible number and diversity of the patterns in his book are a continuous inspiration that can be applied to almost any creative field. There is a strong, incredibly appealing contrast between the organicity and the geometry of the organisms depicted in the illustrations, as can be observed on the following pages.

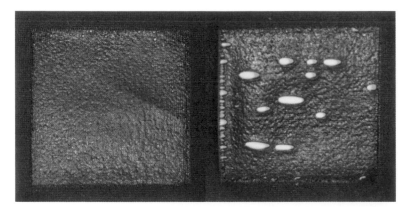

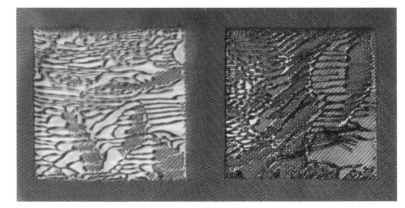

Lycra – from smooth, soft and elastic, to a hard and shiny textile after being exposed to the laser engraving. If exposed for a long time, the material starts melting. Others melt into fine net-shaped structures or become semi-transparent.

Experimentation

The experimentation period is fundamental to obtaining a full understanding of the set of tools of a Fab Lab. Testing what happens to different materials under the same circumstances, for example under the same settings of a machine, is crucial, because it not only allows a better understanding of the material, but also of the limits of the machine itself. These specific limits of the tool are interesting, since every machine leaves some kind of distinctive imprint. Once these limits become clearer, it's easier to convert them into new possibilities and use them to our advantage.

Laser cutter

The research in this project started with the laser cutter, one of the more straightforward machines; the infinite possibilities and combinations this tool alone offers quickly became clear. Every material reacts to the beam of light emitted by the laser in different ways. For example, by engraving certain kinds of synthetic textiles you can change their structure and texture, in a way that makes them almost unrecognizable. Using an archive to document all these samples, and the different options that arose from using different settings and textiles, became an important part of the project. Below, you can see how changing the parameters of speed and power, and the relationship between the two, changes the aspect of the material. Once the research testing the material on its own had been exhausted, I started integrating patterns and different techniques to the process. Old craftsmanship techniques and digital fabrication combine perfectly, complementing each other's weaknesses and strengths. Machines do not entirely replace specific

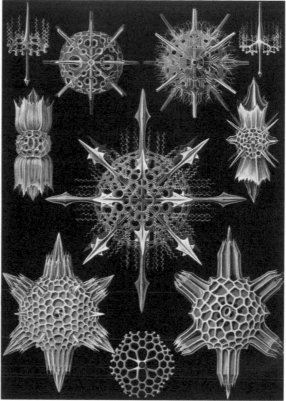

Ernst Haeckel, Kunstformen der Natur,
Tafel 31. Cyrtoidea, Calocyclas.

Ernst Haeckel, Kunstformen der Natur,
Tafel 41. Acanthophracta, Dorataspis.

techniques when manipulating materials, but when blended together with artisanal techniques they can lead to new possibilities and innovation. The equipment of a Fab Lab is a selection of tools that can help us to empower ourselves, making possible the discovery and evolution of techniques that can lead to an infinite number of possibilities.

Starting from the inspiration behind the project, the patterns and techniques that were chosen would have been almost impossible to translate and then realize by hand. However, it turned out that they could be executed perfectly by way of a more digital, machine-based process, such as the laser cutter. The Voronoi-like patterns that Ernst Haeckel studied in different organisms, from the Radiolarians to the sections of different marine invertebrates like corals and algae, are a perfect example.

The digitalization into vectors of different two-dimensional patterns is quite easy and can be translated in different ways. The image on the following page shows the pattern cut out from the textile, creating a lace effect. In the same way, other illustrations pushed me to experiment with other 3D effects, bringing me closer to visualizing and realizing my vision of Ernst Haeckel's drawings. For example, the table showing a series of views of a Limulus (a horseshoe crab), and especially the detail in the construction of their bodies, inspired me to look into braiding and weaving techniques. The underside of these animals resembles a

Laser cut pattern.

The transformation of Ernst Haeckel's
illustration to a vectorial pattern.

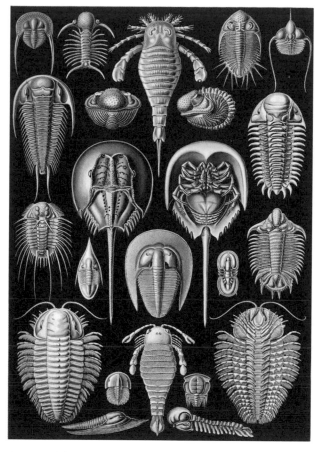

Hernst Haeckel, Kunstformen der Natur,
Tafel 47. Aspidonia.

Visualization by Jens Dyvik and a picture
of the technique executed in rubber.

technique that is commonly used in shoemaking, and of which I found examples throughout the whole Mediterranean area. It consists of a process in which you first cut lines onto a piece of material without separating the strips, and then braid them together using a separate strip of material. This technique changes some properties of the material, allowing it to be deformed, stretched and curved to follow soft curves better. It is characterized by both structural and decorative aspects that reflect the dual concept of the research, as well as the idea of combining craftsmanship and digital fabrication.

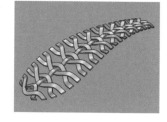

A layering study followed this experiment. A clearer and more 3D effect was achieved through a series of cutout patterns, which, given an offset, would enhance depth perception. Taking once again the Voronoi-lookalike pattern derived from the section study of a coral, then increasing its size and giving an offset of a few millimeters to each layer, the pattern was accentuated by enhancing its volumes and deepening the cutouts. This technique was used in combination with a thick and voluminous textile such as Neoprene to exaggerate this effect. Even the cutouts of this technique are incredibly interesting. At the time of this experiment, the work of my colleague Alessandro Iadarola, who was working on no-waste furniture products, inspired me to look in a different way at what was previously seen as discarded waste. The pieces, sewn together in a specific pattern and order, resemble other illustrations from the book.

Laser cut neoprene textile with 3D effect.

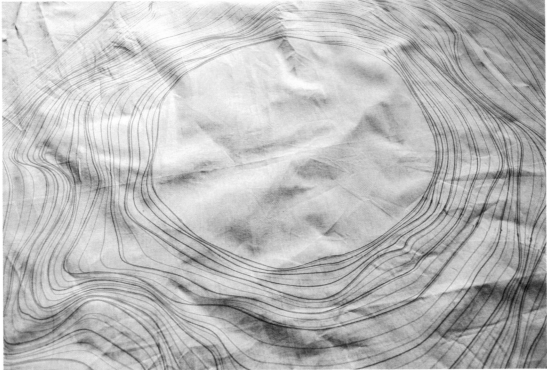

Top: Merel Special pen holder.

Above: drawing on textile with the Merel Special tool.

Opposite page: a picture of the 3D printed ring.

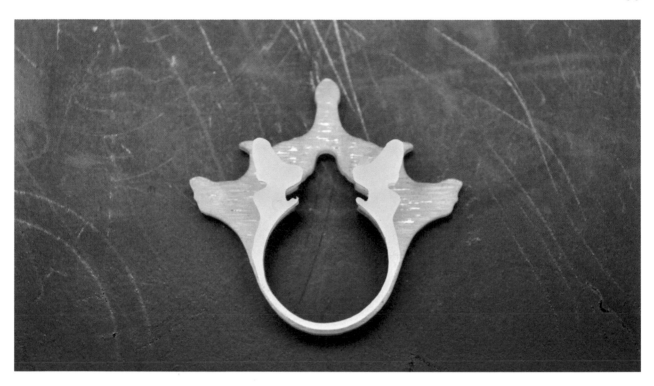

Since the project is entirely focused on fashion, I tried to look at each and every machine not only in terms of what they were intended to do, but more simply in terms of their properties and general functions. In the same way as the vinyl cutter was turned into a sewing pattern printer, the big CNC milling machine was turned into a gigantic drawing table, to imitate a textile printing machine. Printing on small amounts of textile is quite expensive, and is often almost unaffordable for a collection with limited production. Dave Gönner designed a tool for the Shopbot[5] that works as a penholder, the Merel Special; it has a system of elastics that work as shock absorbers, to stabilize the pressure of the pen or marker on the textile, in order to avoid either breaking or damaging its tip. The project is documented on the Fablab Amsterdam website[6]: it comes complete with cutting sheets, which means it can be downloaded, and one can laser cut parts of it and use it in every Fab Lab. The pen is composed of a 6 mm steel rod and a copper tube inserted into the center of the laser cut case made in 3 mm MDF, then assembled with epoxy and wood glue and held together by the elastics to regulate the pressure. With this tool it is possible to draw on any plain surface placed on the table of the milling machine. In this specific case, the textile was secured, using simple double-sided tape, to a wooden board, which was screwed on to the sacrificial layer. By selecting only some lines at the same time, it's possible to change the marker and so to change the color. The result is stunning. The lines are inspired by organic shapes, a collection of soft curves that would have been impossible to draw by hand.

There are certain shapes which cannot be milled in one piece, but are perfect for 3D printing. Inspired by the shape of some corals and shells, I designed, in Rhinoceros 3D[7], a series of lofted shapes that resemble those forms and are perfect for 3D printing. After numerous tryouts of the settings, two different kinds of setting were chosen, one in low quality, where each printed layer is visible and highlights the printed lines effect, and one in higher quality, were this effect is absent and the volumes are more fluid, because each layer is much thinner

Big CNC milling machine

5• Editor's note: the Shopbot is an American brand of big size CNC wood routers; very popular in the Fab Lab network.

6• Gönnor, D., 2013. Drawing attachment for Shopbot | Fablab Amsterdam. Available at: https://web.archive.org/web/20150423050845/http://fablab.waag.org/project/drawing-attachment-shopbot

3D printing

7• Editor's note: Rhinoceros 3D is a proprietary CAD software that focuses on producing mathematically precise representations of curves and freeform surfaces.

Laser cut vertebrae rings.

than in the previous setting. Continuing with some other jewelry designs, the rings in the images are inspired by what is lacking in all the organisms I analyzed in Ernst Haeckel's work: vertebrae. This specific ring resembles the shape of a human vertebrae; a drawing of a lumbar vertebrae was sliced in layers, and these layers digitalized into a vectorial drawing, which was then laser cut in 4 mm acrylic, using different colors for different effects.

Make your own machine

8• Weave – ZigZag Project, 2014. Weave - ZigZag Project – Weaving Europe: Artefacts, Values & Exchanges (WEAVE) - Cycles of cultural events and ateliers to discover Culture and Art through Textiles and Fibre Artefacts in Europe. Available at:https://web.archive.org/ web/20160113100250/http://zigzagproject. eu/

Understanding that each machine in the Fab Lab has its own characteristics, limits and boundaries helps to understand and select the different opportunities that they offer. In addition, the idea of making your own set of tools and fabricating them through the set of machines in a Fab Lab can lead to really interesting developments within the project. As an example, below are shown a set of open design tools that were made using the machines in the lab, to allow the research to expand and progress. This specific set of tools was designed and produced for the event ZIG ZAG, WEAVE project[8] (EU Culture – Weaving Europe: Artifacts, Values & Exchanges; this was a series of cycles of cultural events and ateliers to discover Culture and Art through Textiles and Fiber Artifacts in Europe), organized by Waag Society in collaboration with Museo dei Bambini Società Cooperativa Sociale Onlus (Explora) in Rome, Italy, and the Association Art Land in Sofia, Bulgaria. In addition to these three organizations, seven other European institutions participated: Imaginosity Dublin Children's Museum, Museum of Macedonia, Istanbul Toy Museum, Associazione Le Arti Tessili, Hands

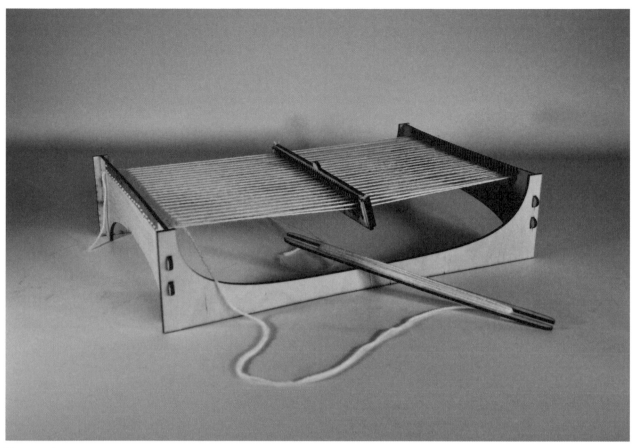

The weaving loom.

On! International, Sofia University St. Kl. Ohridski and Università La Sapienza. The event is specifically aimed at children aged between three and twelve, their families and their teachers. By way of an informal educational approach, they are shown textile-related techniques, to develop their creative skills and learn by doing.

The weaving loom

This small weaving loom is made out of 4 mm plywood with the laser cutter, and has the particular characteristic of not needing to be threaded through the heddle. This means that it's much quicker to start using it, and it offers the possibility of using different selections of heddles. The heddles are held together with elastics, so they are easy to exchange and each one of them selects different threads that create different patterns.

The kite weaving loom

This loom works with the tension provided by the body of the person wearing it. This means that it can only be used when the person is leaning backwards. It has a belt to be strapped to the waist of the user and at the other end attached or hooked to something stable, such as a tree, a strong hook, a pole, or even the leg of a table. This means it can be used inside and outside, and it's easy to carry around.

The knitting loom

This tool is inspired by a Dutch traditional circular knitting tool. It is usually about 10 cm high and has playful shapes, such as a mushroom or a little boy or girl. In this case the

Above: the kite weaving loom at the Zig Zag event.

Right: wooden knitting needles and XXXL knitting needles.

Opposite page: the kite weaving loom at the Zig Zag event.

All the photos in this spread byTomek Dersu Aaron Whitfield www.tomekdersuaaron.com

tools are scaled up in measures that vary from 50 cm wide to more than a meter. It can be used to work on both close circular knitting, giving shape to a tube, or circular open knitting, resulting in a flat piece. It's also possible to decrease and increase the number of stitches.

XXL knitting needles

Two different kinds of knitting needles were created. The first pair has a diameter of 3cm and is made in wood, while the second has a diameter of 10cm, and is created combining two PVC tubes and 3D printed points and caps. The second pair seems to be bigger than the ones in the Guinness Book of Records! This will hopefully be confirmed soon.

Right: bodice basic sewing pattern and application areas of the pattern.
Below: pattern to be applied, sewing pattern and final result of the two combined.

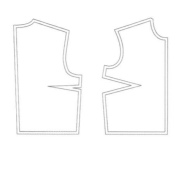

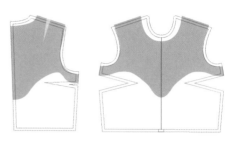

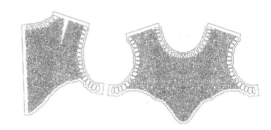

Development

Guidelines and areas highlighted on the sewing pattern of a basic bodice.

Once the first experimentation phase was completed, and the project started taking an actual visual and tactile shape, the results of the studies had to be translated into specific products, in this case a collection of garments and accessories. Working with computer-guided machines means that every aspect of the process has to be converted into digital files, in this case starting from the basic sewing patterns. Software like Rhinoceros 3D, a vector-based 3D modeling tool, makes this conversion quite easy, and without doubt more precise than could be achieved working by hand. In fashion, basic pattern blocks are the start of a sewing pattern, and given specific measurements of a body, it's incredibly easy and it takes much less time, space and material to draw them on the computer.

The digital sewing patterns combined with a tool like the vinyl cutter – standard inventory of a Fab Lab – give you immediate feedback on your work. The vinyl cutter, which is usually used to cut out stickers, can also be used to print out digital drawings, since it is computer-guided and follows the vector-based lines of the drawing. In this project it was used to print out the outlines of the sewing patterns that I had digitalized. This procedure enables one actually to check and test the patterns, and when there are changes to be made, it's quick and easy to modify the digital files, rather than redrawing or modifying the pattern by hand. In addition we can change the measurements, proportions and size of the garment. After testing the basic patterns, the real design process starts, in which the actual shapes and overall look of the collection have to be defined.

This collection reflected its inspiration by contrasting the clean cuts made by the laser cutter, which echoed the scientific approach of Ernst Haeckel's research, with the soft, light bottom parts of the dresses, which echoed the beautiful, dreamy and charming watercolors through which he explained his findings. Soft and hard materials and shapes are combined together to express the contradiction of the two conflicting worlds of artist and scientist. At this point in the experimentation phase, the results need to be translated and applied to the design of the garments. One also has to keep in mind that the results of this phase also need to respond to the structural needs of the designs, since the manipulation of the material itself can change its structure and properties.

Illustrations of the collection;
black ink on paper.

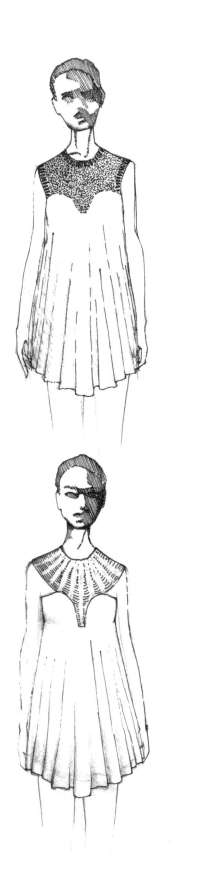

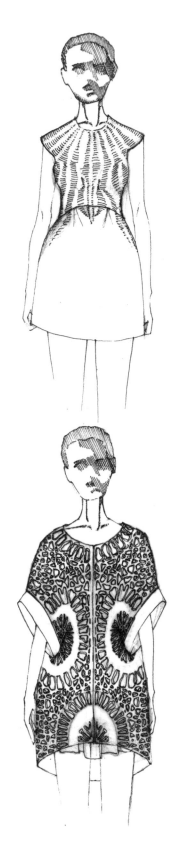

Pictures of the collection at
Makers festival Amsterdam;
photos by Alex Schaub,
model Josephine Keuter.

The coat, from the basic sewing pattern
to the final cut out pattern.

Going back to the illustrations in Artforms in Nature, I defined the areas and the way in which I wanted to intervene on the first dress, and selected which pattern I wanted to apply. The experiment which I had previously tested on the laser cutter was modified, repositioned and applied to the shape of the upper part of the bodice of the dress in question, as you can see in the images below. The pattern has been redrawn and positioned in a way that defines the outline of the dress, highlighting its shape. If the textile – for example, lace that resembles the cutouts – had been bought in a shop, this wouldn't have been possible, because the positioning of the pattern gives much more detail to the dress, and shows it has been designed in this specific way. In this way the material reflects the needs of the design and better represents the idea behind it. Once the file is ready, it's simply transferred to the computer connected to the laser cutter and immediately tested on a material similar to the one to be used for the final product. Again, the testing is crucially important. Small mistakes and changes can be adapted in the drawing in a matter of minutes; if the pieces had been hand-cut, this process would have been extremely long and all but impossible. The bottom part of the dress was also digitalized and printed with the vinyl cutter on tracing paper, and then cut by hand, since it didn't have any particular cutouts or details. This was then assembled with a simple sewing machine and refined by a clean, thin overlock stitch.

The second technique I wanted to experiment with was the braiding pattern, which I decided to apply to a more complex dress. This dress took much more work and time, since the technique itself changes the properties of the textile that is used. Starting from the basic sewing pattern of a bodice, main areas and guidelines were established. Being a dress with a fitted bodice, darts also had to be incorporated into the braiding pattern. This creates a further challenge but also shows, once again, that the properties of this technique contribute not only in a decorative but also a structural way. Given the more complex shape of the dress and hence the technique, the process took much longer than the other pieces. From a single-pieced body, the dress evolved into a two-piece for the front and four-piece for the back. The bodice and the skirt are braided together and don't actually have to be sewn, because the cut-out lines connect to create a one-piece effect. The integration of the hand-braided technique with the clean-cut lines of the laser cutter reflect, once more, the hard and soft, scientific and

Opposite page and right:
bag details.

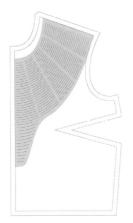

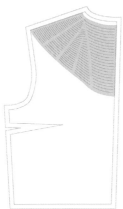

Highlighted section that defines the
collar of the third dress.

artistic contradictions that Ernst Haeckel highlight in his book. Accentuating the combination of handcraft and machine work creates a more perfect and complete result.

The third dress is a combination of the two previous ones, in which we find a decorative collar shape similar to the first dress and its light, flowing body, together with the technique used in the second dress. The experiment with the layering of textile was applied to the coat, where the layers accentuate the silhouette of the garment and harden its appearance. The sewing pattern was digitalized and tested following the same procedure used when testing the dresses. Following this, I adapted the coral section pattern to the sewing pattern of the coat, in order to bring out its curved details, especially around the collar for the front and the shoulder for the back. Then the pattern was cut out in three different layers of Neoprene, comprising three separate coats, which were then sewn together. More details about the offset and placement of the pattern can be seen in the images below. The bottom layer and lining of the garment shows through the cutouts, highlighting the sense of depth, which was the main aim of using this effect. The lining of the coat is a thick cotton canvas and was hand-dyed to obtain the exact color I was looking for. The leftovers of this process were used to cover the outer shell of a bag. The concept behind the bag is mostly functional, reflecting my personal needs in terms of this specific type of item: it is a small bag without any kind of visible strap or handle that is easy to carry around and to cycle with. It has a hidden pocket covered by the cutouts, through which you can stick your arm, to be able either to hold the other side of the pocket in your hand, or to slip your hand through it to be able to do whatever you wish, leaving your hands free. Every single shape is then sewn onto this layer and to the rest of the outer part of the bag, in order to hide the entrance of the pocket and decorate the bag with a look similar to some of the illustrations of my inspiration.

This small collection represents the start of my research into the relationship between digital fabrication and fashion design, which will be extended through new projects and collections. The following pictures were taken at the Makers Festival Amsterdam. During this event, the visitors were invited to try on the garments.

Massimo Menichinelli

Fab Gallery

Projects from and for Fab Labs

Each Fab Lab has its own history, expertise, knowledge and equipment. Each Fab Lab has its own community, with different user profiles, needs and activities. This means that many different kinds of projects can be developed in Fab Labs, according to the specific context and to their resources in terms of time, money and expertise. Projects can range from simple exercises or prototypes to finished products. This chapter sets out many examples of what can be achieved in a Fab Lab, or with the help of a Fab Lab, or even inspired by a Fab Lab, even if the project is not realized within the Fab Lab network.

Projects: visual index

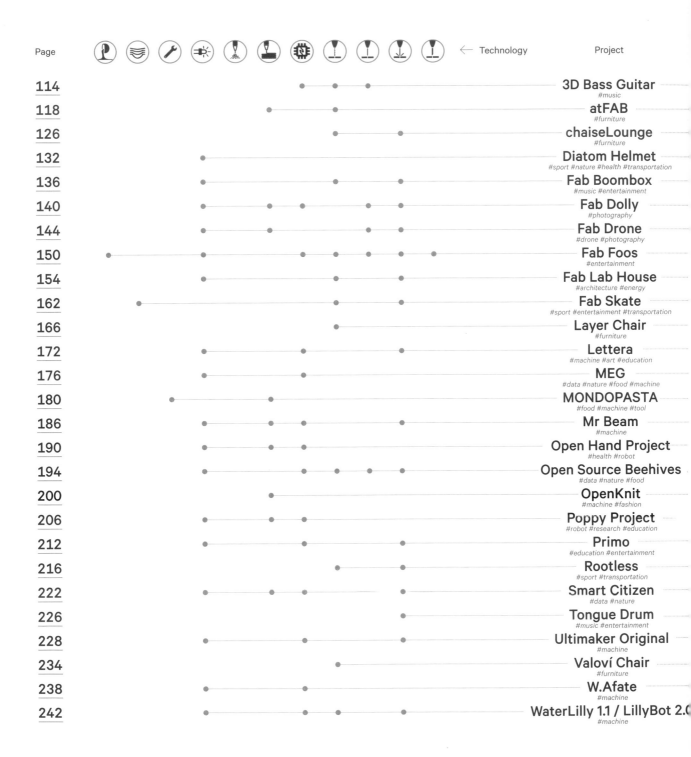

Places

EUROPE

Finland

Norway

UK

The Netherlands

France

Germany

Switzerland

Italy

Spain

Canada

USA

China

Japan

Togo

Indonesia

Brazil

New Zealand

AUTHOR.

Alex Schaub
Carsten Lemme
Alice Mela
Thomas van der Werff

DATE.
2011

DESIGNED/FABRICATED.
FABLAB AMSTERDAM
(AMSTERDAM, THE NETHERLANDS)

TECHNOLOGIES USED.
BIG CNC WOOD ROUTER
CNC MILLING
ELECTRONICS PRODUCTIONS
MICROCONTROLLERS

LINK.
HTTPS://WEB.ARCHIVE.ORG/
WEB/20160112044632/HTTP://FABLAB.WAAG.
ORG/PROJECT/3D-BASS-GUITAR

PHOTOGRAPHY.
ALEX SCHAUB / CARSTEN LEMME / ALICE MELA
THOMAS VAN DER WERFF / AURÉLIE GHALIM

PROJECT.
3D Bass Guitar

#MUSIC

A widely held misconception about Fab Labs and digital fabrication is that their principal use is the production of prototypes. In fact, many digital fabrication technologies are used for building finished products, and many finished products are also created in Fab Labs. It is generally a question of time, money and experience: the casual user doesn't have enough of any of these, and will probably make few prototypes. But if we invest time and resources in learning how to work and develop projects in a Fab Lab, anyone is capable of producing astonishing projects. An example is Alex Schaub's 3D Bass Guitar, a fully functional bass guitar made in its entirety at Fablab Amsterdam.

The process was based on the use of a large CNC wood router: starting first from a foam model for a prototype, and ending with hard wood for the finished object. The object was CNC milled on two sides, first the body and then the neck. The fine texture made it a very long process (more than 45 hours); the body of the bass was CNC milled with a custom 3D pattern, which also contributed to the length of the project. Pickups and other electronics were also custom-made in the Fab Lab, but with an old (and not CNC) milling machine. The 3D Bass Guitar works as perfectly as a handmade bass guitar created by an expert.

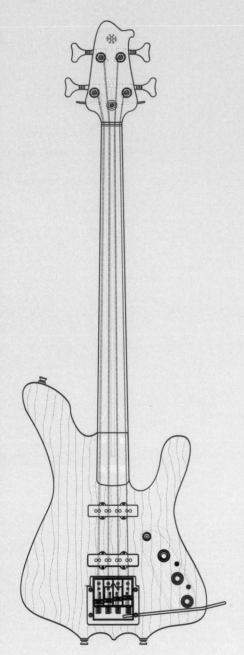

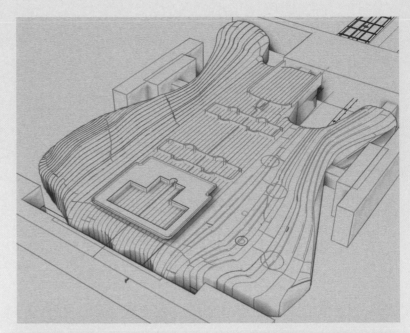

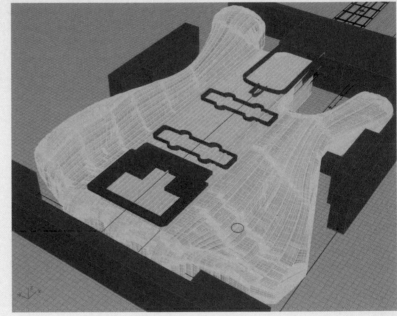

The 3D Bass Guitar has a complex 3D shape, so it had to be 3D modeled in Rhinoceros and then 3D milled with a big CNC wood router. The process required a lot of handwork, especially in the finishing of the wood surface and the assembly of all the electronic components.

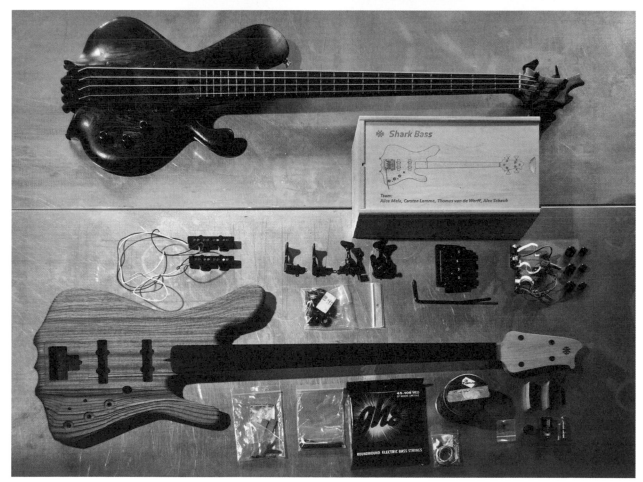

AUTHOR.

Anne Filson
Gary Rohrbacher

PROJECT.

atFAB

#FURNITURE

DATE.
2012

DESIGNED/FABRICATED.
DESIGNED IN THE USA
FABRICATED GLOBALLY

TECHNOLOGIES USED.
BIG CNC WOOD ROUTER
3D PRINTING:
FUSED DEPOSITION MODELING (FDM)

LINK.
HTTP://ATFAB.CO/
HTTPS://WWW.OPENDESK.CC/ATFAB
HTTP://WWW.THINGIVERSE.COM/ATFAB/DESIGNS
HTTP://FILSON-ROHRBACHER.COM

atFAB, a design firm, was co-founded by architects Anne Filson and Gary Rohrbacher. All of the models are open design projects that may be downloaded and fabricated locally. The projects were designed with technology commonly found in Fab Labs and other workshops, while the designs are optimized for big CNC wood routers and plywood, a very popular material in Fab Labs. The principal goal of the project is the building of a series of objects that can be manufactured locally by a global community of Makers. The production process is intended to be more environmentally friendly, as well as more accessible for the Fab Labs and small, local and independent manufacturing companies. The objects in atFAB's collection also feature smaller components that can be downloaded and 3D printed.

atFAB leverages the power of Opendesk, an online platform for the sharing of open design projects, in particular ones that can be manufactured with big CNC wood routers. The projects were designed in the USA, but they have been fabricated in many different countries. This is a great example of how professional designers are starting to respond to distributed networks of Fab Labs and similar workshops in the production of their projects.

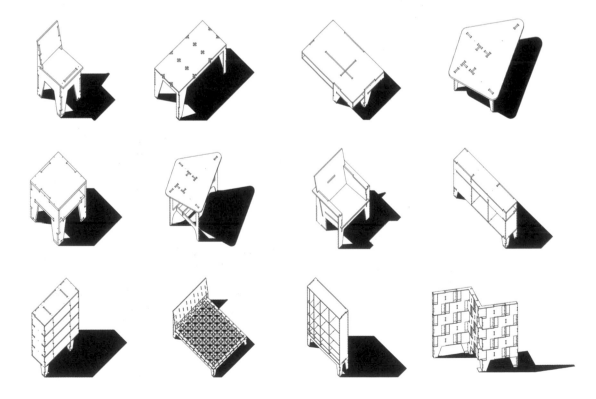

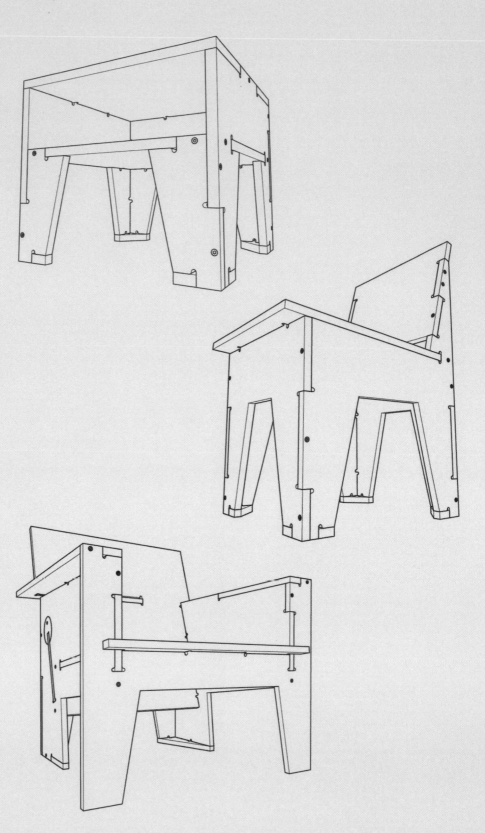

Beside Table

Designed as a side table with a storage shelf, the Beside Table is comprised of 11 interlocking pieces. They resolve loads and forces with symmetrically arranged tapered leg assemblies that counteract shear forces. Two tables can be constructed from a single 4' x 8' sheet of material. The table is designed to be transformed dimensionally in height, width and depth to serve different contexts and needs.

Five to Thirty Minute Chair

The Five to Thirty Minute Chair is a multi-purpose side chair, which can be made of almost any material and finished as desired. Two chairs can be milled from a single 4' x 8' sheet, with each one comprised of ten flat, interlocking pieces that are easily constructed and secured with screws, pegs or adhesives. It can also be accessorized with a 3D printed Peg & Foot Kit.

The Ninety Minute Chair

is a lounge chair comprised of nine pieces. Loads are distributed across its interlocking seat, arms, and legs. It can be made of almost any material and finished as desired. As shown, one chair can be milled from two sheets of 4' x 8' material. The chair requires a set of 3D printed keys that simplify assembly and lock pieces into place.

One to Several Table
Designed for working, dining and meeting, the One to Several Table is comprised of 13 interlocking pieces. It is a lightweight and stable structure that relies on a torsion box top and a rotationally symmetrical arrangement of diagonal legs. It is designed to be dimensionally transformed in width, depth and height with a parametric app. The design accomodates optional grommets in its top to manage wires.

The **Rotational Table** is a freestanding table with cantilevered corners and storage compartments that are accessible on all sides. Comprised of ten interlocking pieces, its loads and forces are resolved by a rotationally symmetrical arrangement of tapered leg assemblies. The Rotational Table is designed so it can be dimensionally transformed in height, width, depth and compartment size. The cut files include one 680 x 680 x 680 table that can be cut from a single sheet of 4' x 8' material.

The **Rotational Stools** are a complementary pair that share common proportions (15" tall x 9.5" square) and variations on a rotational structure. They function equally well as low stools or as side tables – individually, in pairs, or in aggregate. A pair of Rotational Stools requires a quarter of a sheet of plywood. You can make eight stools from a single sheet of plywood, or nest parts into unused areas of plywood when you're cutting other projects.

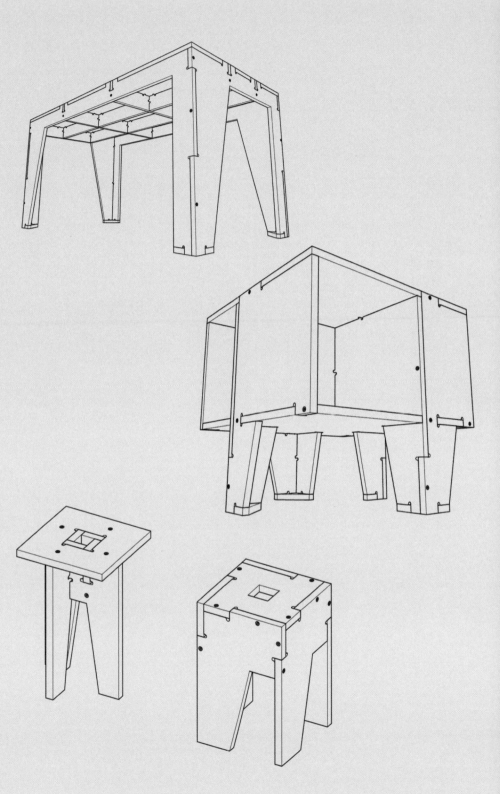

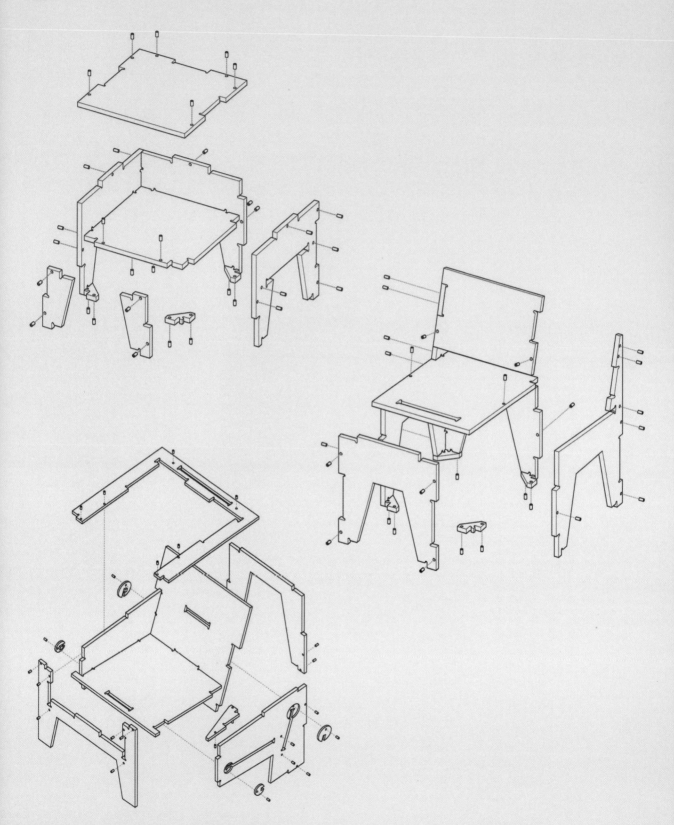

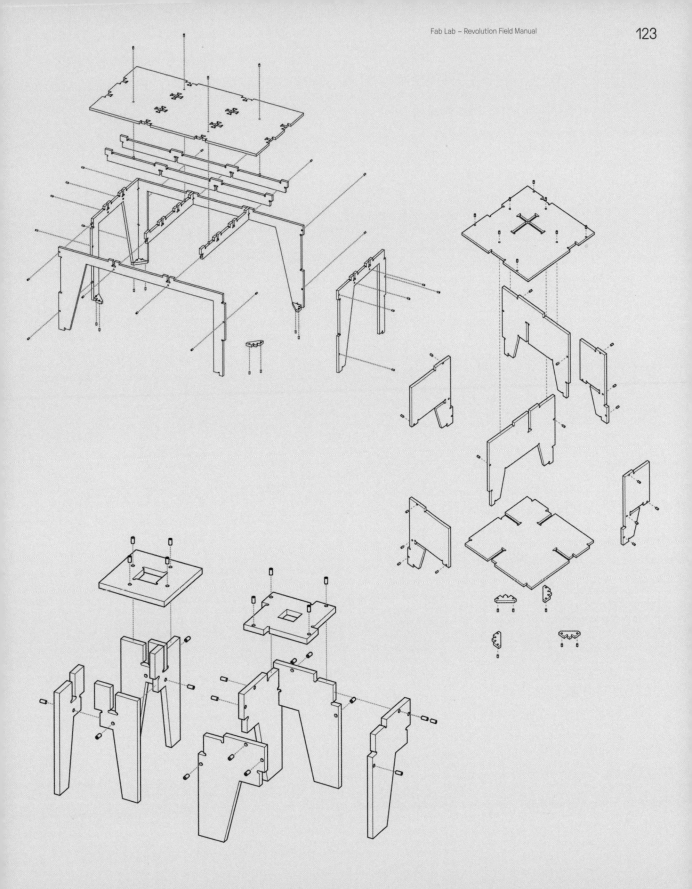

Rotational Table

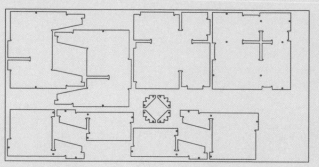

Beside Table

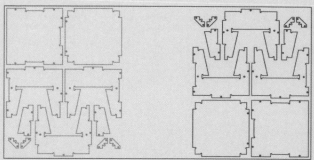

Rotational Stools

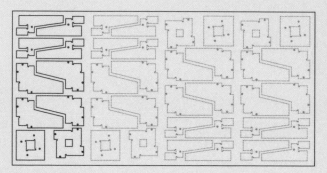

Five to Thirty Minute Chair

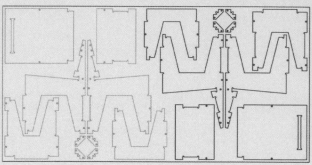

One to Several Table

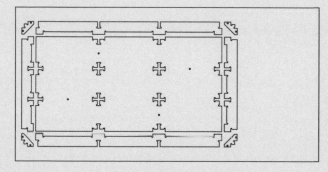

The Ninety Minute Chair

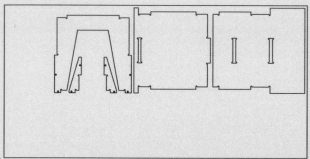

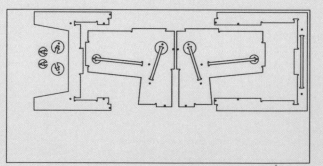

MUCH FROM LITTLE

A LINE OF FURNITURE
FROM A SINGLE DETAIL

EXISTING FLOWS & NETWORKS

CNC
MACHINES

SHEET
MATERIAL

FABLAB & FAB
NETWORKS

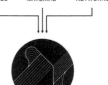

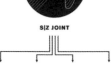

S|Z JOINT

FURNITURE OBJECTS

SHIP INFO NOT STUFF

LESS TRANSPORT
LESS ENERGY

DIGITAL CUT FILES

BUILDS BY MAKERS WORLDWIDE

INFINITE VERSIONS

FURNITURE ANYWAY
ANYPLACE ANYTIME

PARAMETRIC DEFINITIONS

DIMENSIONS

MODULE

SHAPE

MATERIAL
THICKNESS

BIT/BEAM
DIAMETER

CUSTOM
DETAILS

CUSTOMIZE
WITH
ONLINE
APP

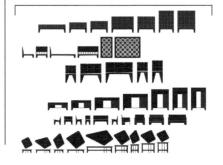

AtFAB, which stands for 'FABrication at', has built its business on the customization and fabrication of its products. Users can download the designs from atFAB's page on Opendesk, an open design platform which allows everyone to see where the projects have been downloaded and fabricated worldwide. Projects can be then modified by designers and people who know how to design with CAD tools. But in order to fully democratize these projects, atFAB

also built an online platform, based on the open source programming language Processing, which enables anyone to change the parameters of some of its projects. Shapes, dimensions, thicknesses and even CNC tollbooths can be fully customized online. This is a great example of a meta-design approach: the conscious construction by the designers of the same tools and processes that will be used in the whole lifecycle of the design projects. With this emerging

approach, even non-professional designers or users without design and CAD skills can have a role in adapting each project to their own local needs. But for users who do not live close to a Fab Lab, a makerspace or a fabrication facility, atFAB also runs its own online shop for selling furniture that has already been fabricated. The users will still have to assemble this furniture, and atFAB also ships the instructions and the required hand tools for the assembly.

AUTHOR,

Pietro Leoni

PROJECT,
chaiseLounge

#FURNITURE

DATE,
2012

DESIGNED/FABRICATED,
DESIGNED AND FABRICATED AT FABLAB TORINO
(TURIN, ITALY); FABRICATED AT AALTO FABLAB
(HELSINKI, FINLAND)

TECHNOLOGIES USED,
BIG CNC WOOD ROUTER
LASER CUTTING

LINK,
HTTP://PIETROLEONI.COM/#CHAISELOUNGE

PHOTOGRAPHY,
PIETRO LEONI
AALTO FABLAB

Over the last 15 years, many designers, researchers and activists have proposed the idea of sharing the blueprints and instructions of design projects as part of the open design philosophy. As a further sign of how the open design philosophy has been emerging and forming part of the mainstream, in 2012 *Domus* magazine launched a competition for design projects that could be manufactured in Fab Labs. Following the competition, these projects were published as open design. The chaiseLounge is one of the outstanding projects in the competition: designed by Pietro Leoni, it comprises a small hammock, built on a single sheet of 10 mm plywood that can be milled with a big

CNC wood router. The design consists of a series of triangular elements that interconnect to form a kind of wood membrane, which adapts to the user's body. The project, initially developed at Fablab Torino, was also the basis for a collaboration with the Aalto Fablab in Helsinki, where a modified version was laser cut for an exhibition during the Helsinki World Design Capital 2012. In this way, the object was produced locally and adapted to the locally available materials and machines, rather than being shipped from Turin to Helsinki: another example of the flexibility to which the adoption of the open design philosophy can give rise.

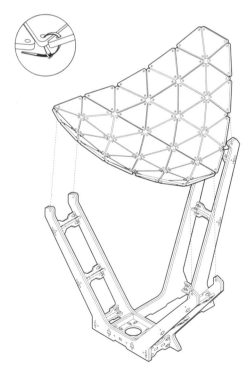

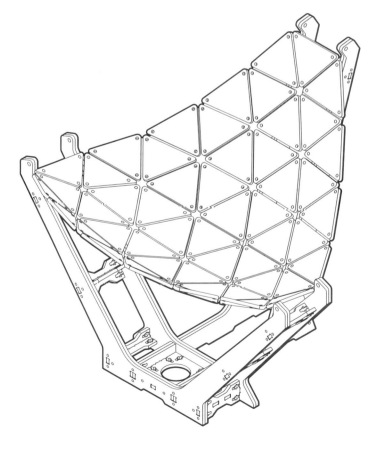

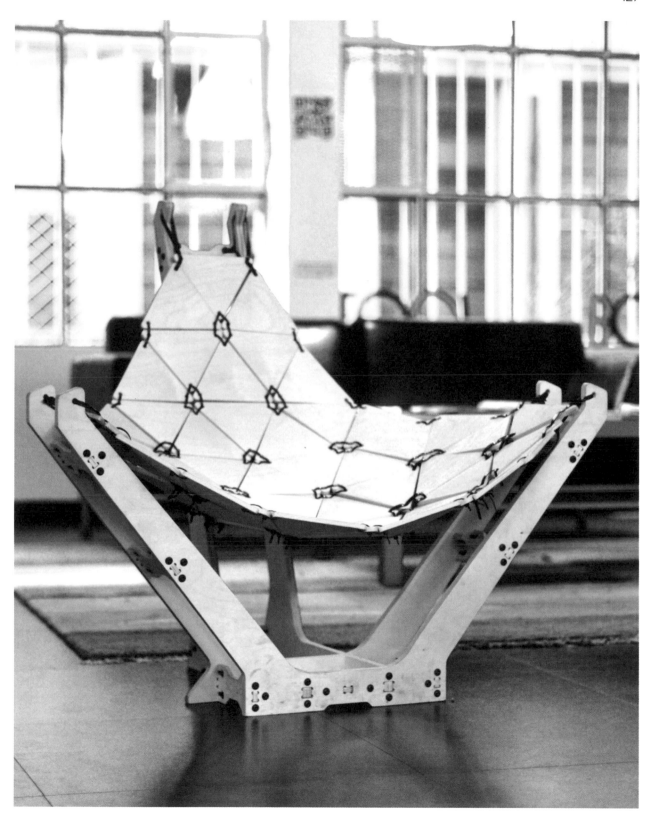

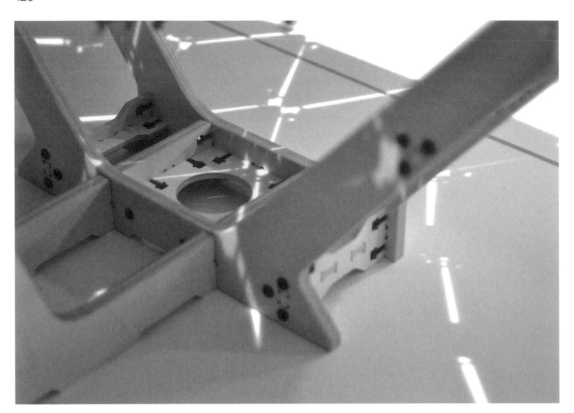

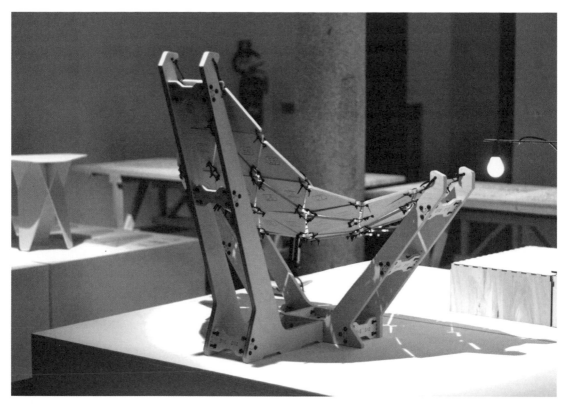

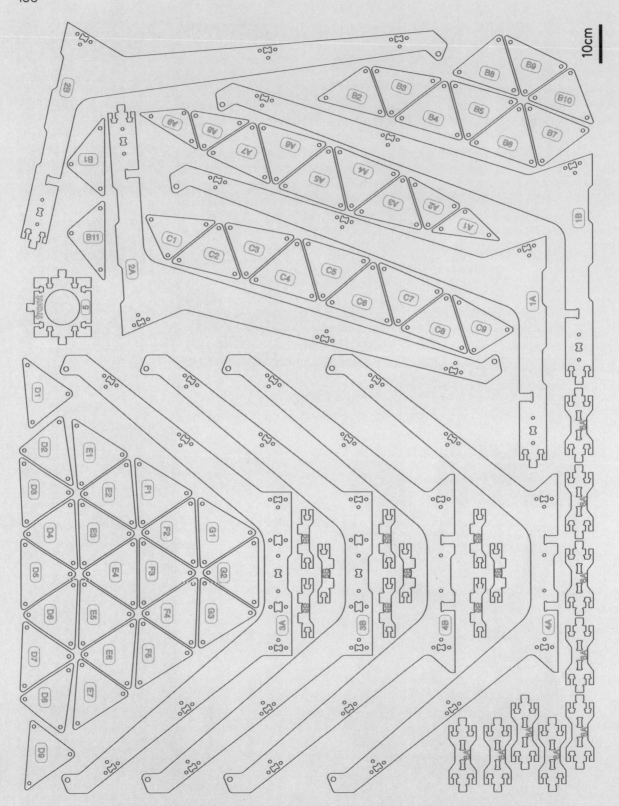

10cm

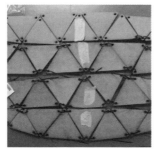

The Helsinki version of the chaise-Lounge was laser cut instead of CNC milled with a big CNC wood router, since this machine wasn't available in the Fab Lab at the time. Pietro Leoni then designed a custom version for the laser cutter, building it from more layers of thinner material.

The laser cut version is somewhat different to the CNC milled version, since a different process (burning instead of cutting) and a different material (MDF board instead of wood) were used. This is a common feature of many of the projects manufactured in a Fab Lab: it is easier to spot the manufacturing processes adopted in the fabrication.

AUTHOR.
Paula Studio
Antonio Gagliardi

DATE.
2014

DESIGNED/FABRICATED.
PAULA STUDIO (ROME, ITALY)
MAKEINBO (BOLOGNA, ITALY)

TECHNOLOGIES USED.
3D PRINTING: INKJET POWDER PRINTING (3DP)

LINK.
HTTP://WWW.ALLABOUTPAULA.COM/WORKS/
DIATOM-HELMET/
HTTP://DIATOMDESCIENCE.TUMBLR.COM/

NOTES.
CONCEPT AND DESIGN:
PAULA STUDIO (SIMONE BARTOLUCCI,
VALERIO CIAMPICACIGLI)
PARAMETRIC DESIGN:
ANTONIO GAGLIARDI (MAKEINBO)

PROJECT.
Diatom Helmet

#SPORT #NATURE #HEALTH
#TRANSPORTATION

Artists, architects and designers have always taken inspiration from nature. Words like biomimetics, bionics and biomimicry define models, systems, and elements drawn from nature that are used to solve complex human problems. In the previous chapter we saw how the aesthetics of living organisms can interact with the aesthetics of design projects developed in Fab Labs; sometimes, we can also derive ideas about structural issues for working in a Fab Lab. In this regard, the Diatom Helmet is a very interesting project, having being developed at the intersection of nature, design and science. It is a collaboration between a professional design studio (Paula Design) and the Fab Lab of Bologna (MakeInBo). This project was developed for a call for the DIATOM De-science exhibition in July 2014 (at

Città della Scienza, the Science City in Naples, Italy). Curated by Prof. Carla Langella, it was organized into three research units: a biology, a physics, and a biomimicry design research unit at the Seconda Università degli Studi di Napoli, and a micro- and nano-electronics research unit at an institute (IMM CNR). The exhibition focused on diatoms, a major group of algae and one of the most common types of phytoplankton. These algae are famous for being enclosed within a cell wall made of silica, known as a frustule. The holes in the frustule provide high resistance to compression; it would otherwise be very difficult to have a structure made of silica. The key to the frustule's strength is its atoms. We can thus learn from diatoms how to build very strong structures using very little material, in a highly efficient and resource-cheap process.

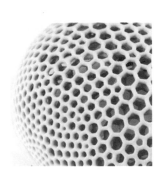

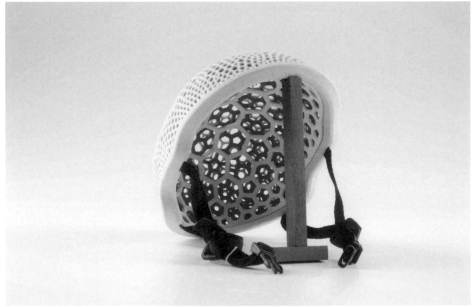

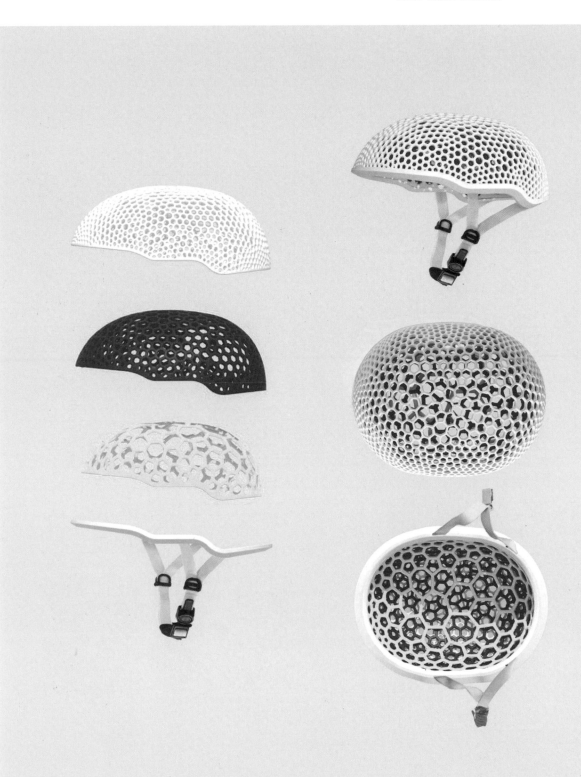

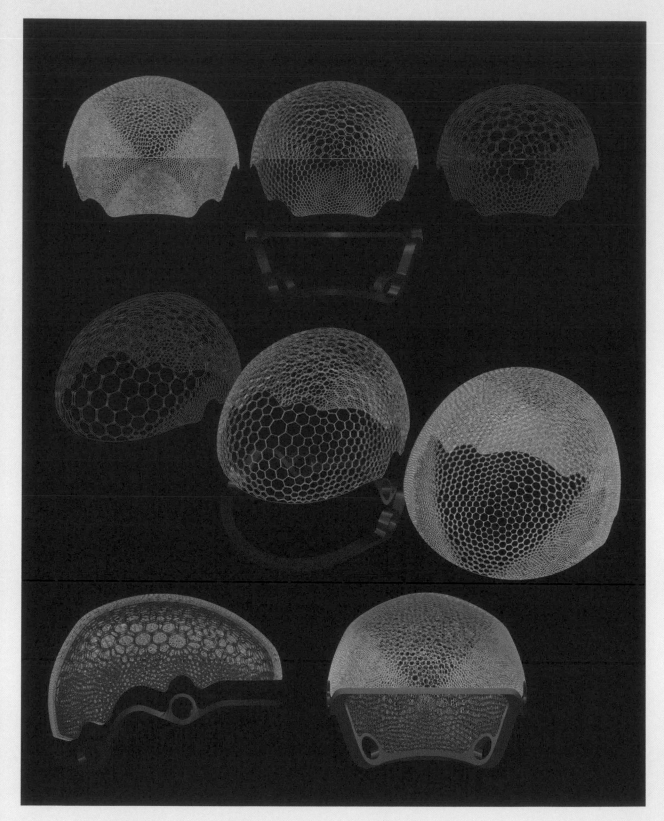

The helmet was designed with this structure in mind. It is made up of three different layers, with parametric patterns of holes. The first prototype was 3D printed with an Inkjet Powder Print machine, and later hand painted and assembled with other materials and components; the final product will be mass produced using different materials and processes. The role of the Fab Lab in this project was, unexpectedly, not that of hosting the 3D printing process (which was undertaken outside of the lab) but of providing knowledge and expertise in designing the parametric structure. This shows how Fab Labs are important not only for providing access to technology, but also for giving access to knowledge and collaborators. Even small Fab Labs can participate actively in the development of innovative projects.

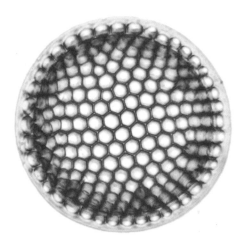

Diatom (fossile) – 630x by Michael Fingerle, aka Picturepest.

AUTHOR.

Matthew Keeter

PROJECT.
Fab Boombox

#MUSIC #ENTERTAINMENT

DATE.
2011

DESIGNED/FABRICATED.
CENTER FOR BITS AND ATOMS – MIT
(CAMBRIDGE, USA)

TECHNOLOGIES USED.
LASER CUTTING
CNC MILLING
ELECTRONICS PRODUCTION

LINK.
HTTP://FAB.CBA.MIT.EDU/CLASSES/863.11/
PEOPLE/MATTHEW.KEETER/FAB_BOOMBOX/
HTTP://FAB.CBA.MIT.EDU/CONTENT/PROJ-
ECTS/BOOMBOX/
HTTP://WWW.MATTKEETER.COM/PROJECTS/
BOOMBOX/

PHOTOS.
MATTHEW KEETER

Fab Labs and the Center for Bits and Atoms were founded at MIT after Neil Gershenfeld's 'How to Make (Almost) Anything' course. This course provided a conceptual basis for the Fab Lab concept, and was extended repeatedly both at MIT and globally, through the Fab Academy education format, which enables the course to be offered within a network of Fab Labs.

One of the most successful and well-known projects derived from this course is the Fab Boombox, a custom project designed using the technologies commonly available in a Fab Lab. Designer Matthew Keeter concentrated on simplicity: it can be fabricated for less than $100, it uses a single 9V battery, and MP3 files are stored in a standard SD card; it has capacitive buttons, engraved on the plywood, that respond to simple finger touches; and it was designed as a parametric project, where the choice of different values for the design parameter of the project can generate boomboxes of different sizes.

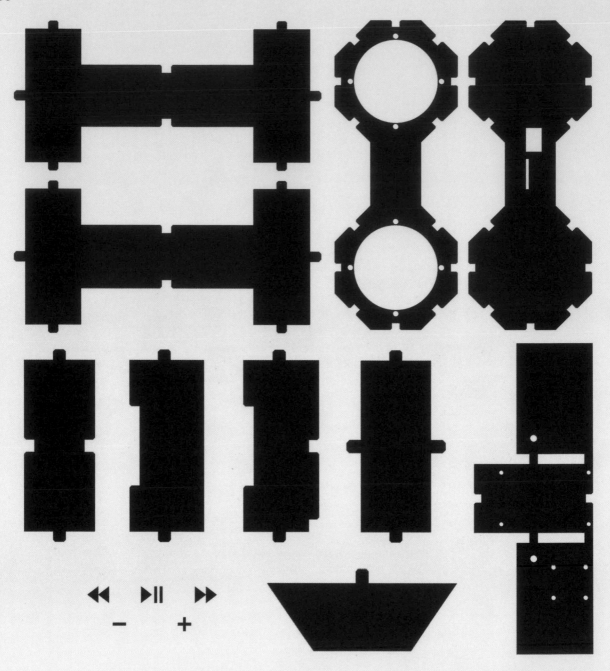

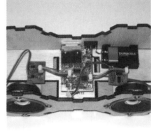

The case of the Fab Boombox was laser cut and the keys were laser engraved on it. The design is generated by custom software that enables parameters to be tweaked.

The Fab Boombox is a mix of industrially fabricated components and of digitally fabricated custom components.

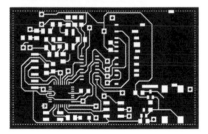
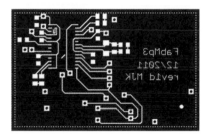

The main board of the Fab Boombox consists of a two-side CNC milled PCB. This is a more complex process than a single side board, but it is still possible in a Fab Lab.

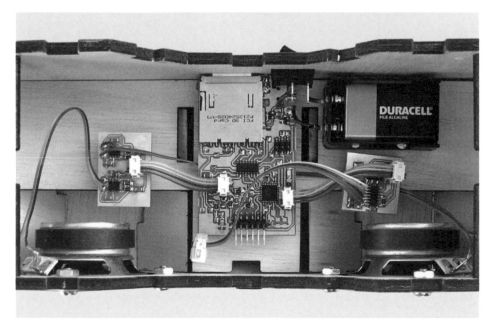

The inside of the Fab Boombox shows how the project itself is quite simple in terms of number of components. Many Fab Lab projects are accessible not only because the source files can be downloaded, but also because they are designed to be fabricated with few and cheap components.

AUTHOR.

Moritz Begle

PROJECT.
Fab Dolly

#PHOTOGRAPHY

DATE.
2014

DESIGNED/FABRICATED.
FAB LAB BARCELONA (BARCELONA, SPAIN)

TECHNOLOGIES USED.
LASER CUTTING
CNC MILLING
3D PRINTING: FUSED DEPOSITION MODELING
ELECTRONICS PRODUCTION
MICROCONTROLLERS

LINK.
HTTP://FABACADEMY.ORG/ARCHIVES/2014/
STUDENTS/BEGLE.MORITZ/

PHOTOS.
MORITZ BEGLE

Each Fab Lab offers its own range of courses and workshops, but there is one educational format that is increasingly widespread, namely, the Fab Academy. The Fab Academy, a concept derived from the above mentioned 'How to Make (Almost) Anything' course, has become the principal educational format in the world of Fab Labs. This initiative has contributed to the expanding collaboration between established and emerging Fab Labs. Students work on a series of exercises and a final project, which often are very representative of the kind of projects currently being developed in Fab Labs. A very interesting project from the Fab Academy of 2014 is the Fab Dolly, an open source camera dolly that can be fabricated in a Fab Lab.

A camera dolly is a specialized bit of equipment used in photography and filmmaking: it is designed and used to enable smooth camera movements, by moving the camera along a horizontal track. In response to the fact that camera dollies are typically very expensive, Begle Moritz designed and fabricated an open source and low-cost version that can be used with cameras and 3D laser scanners. This project is an automated camera dolly system that is easy to create and use, and is adaptable to different cameras. It can also be used to move along three axes. Instead of traditional steel and aluminum (strong but expensive), Begle Moritz chose 5 mm laser cut plexiglass (PMMA), very popular among Fab Labs; the gears and wheels were 3D printed with an FDM 3D printer. An Arduino controls the movements of the Fab Dolly, which can be controlled from a desktop application built with Processing.

Details of the gears and of the main mechanism of the Fab Dolly, designed and fabricated from scratch in the Fab Lab.

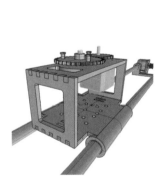

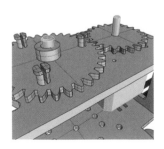

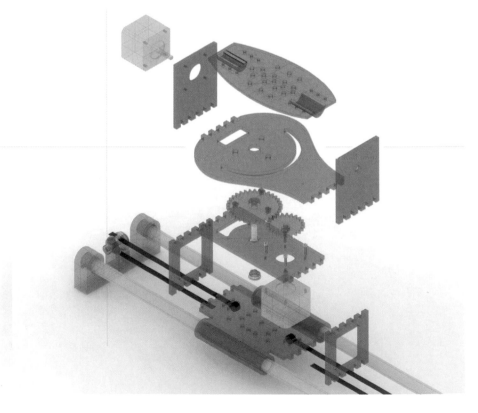

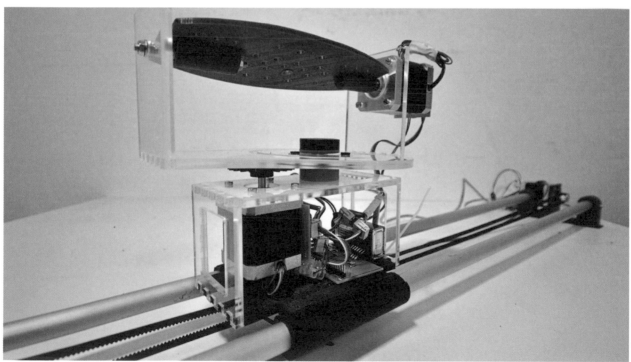

3D Printer:

Laser cutter:

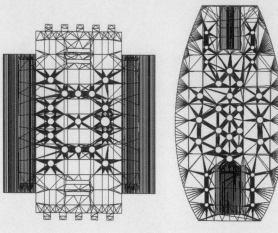

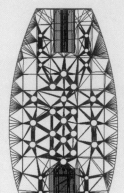

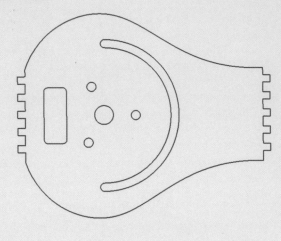

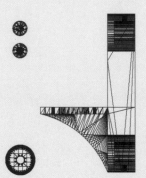

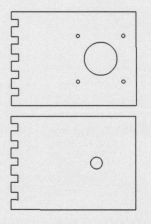

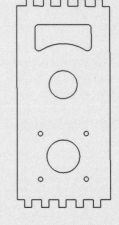

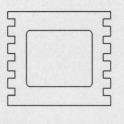

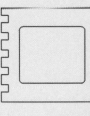

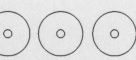

The manufacturing processes behind Fab Dolly were quite common and simple. The complexity of this project is in the mechanism design.

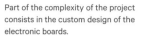

Part of the complexity of the project consists in the custom design of the electronic boards.

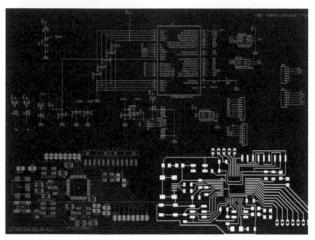

AUTHOR.

Chirag Rangholia
Aldo Sollazzo

PROJECT.
Fab Drone

#DRONE #PHOTOGRAPHY

DATE.
2014

DESIGNED/FABRICATED.
FAB LAB BARCELONA (BARCELONA, SPAIN)

TECHNOLOGIES USED.
LASER CUTTING
CNC MILLING
ELECTRONICS PRODUCTION
3D PRINTING: FUSED DEPOSITION MODELING
(FDM)

LINK.
HTTP://FABACADEMY.ORG/ARCHIVES/2014/
STUDENTS/RANGHOLIA.CHIRAG/
HTTP://FABACADEMY.ORG/ARCHIVES/2014/
STUDENTS/SOLLAZZO.ALDO/

PHOTOS.
CHIRAG RANGHOLIA
ALDO SOLLAZZO

Drones, also called unmanned aerial vehicles (UAV), are, alongside 3D printers, the most well-known example of a product associated with the Maker movement. There are many types of drones, from highly complex military drones to very simple DIY models. The most common model is the quadcopter, a (small) multirotor helicopter that is lifted and propelled by four rotors. Quadcopters can be found in many Fab Labs, and the Fab Academy is no exception. The Fab Drone project is a very interesting example of how we can design and build a quadcopter with the materials and technologies available in Fab Labs. Developed collaboratively by Chirag Rangholia and Aldo Sollazzo, Fab Drone is a low-cost and open source quadcopter. The frame of the quadcopter is made of laser cut plywood, generated with a parametric project that can create many different designs of drones. It can be used with different materials, motors, propellers and electronics boards. Furthermore, using cheap plywood makes it easy, quick and cheap to recreate the frame following crashes, getting it into the air again in a very short space of time.

The main electronic board is an inertial measurement unit (IMU), an electronic device that measures the quadcopter's velocity, orientation, and gravitational force. It uses a series of sensors which include accelerometers and gyroscopes. These sensors, in combination with the data they gather and the onboard algorithm that filters it, stabilize the drone. The IMU of this project, fabIMU, is based on the open source ArduIMU board, a custom Arduino-compatible board designed and manufactured specifically for drones. Fab Drone also has a 3D printed and motor controlled gimbal, a pivoted support that allows the rotation of an object, which can carry and rotate a GoPro camera. The drone can be controlled with an Android mobile app taken from the Flone project.

The gimbal that holds and controls the camera is attached under the Fab Drone.

The Fab Drone mostly consists of
custom digitally fabricated components,
with few industrially manufactured ones.

The brain of the Fab Drone consists
of a digitally fabricated main board
plus some industrial boards. It is a very
complicated project.

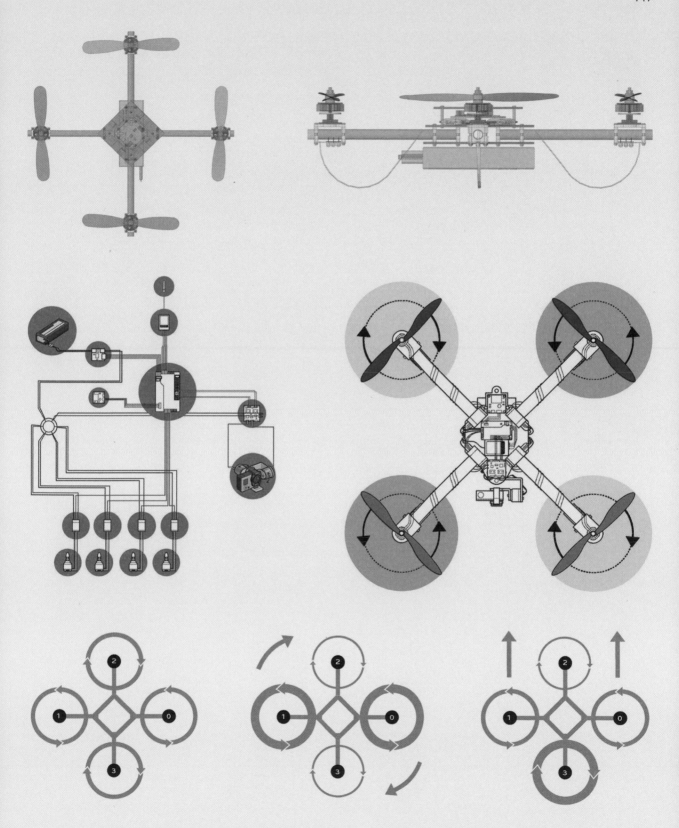

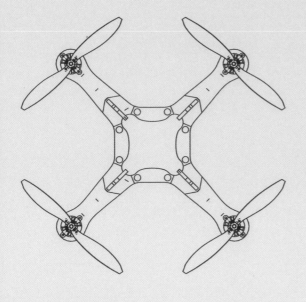

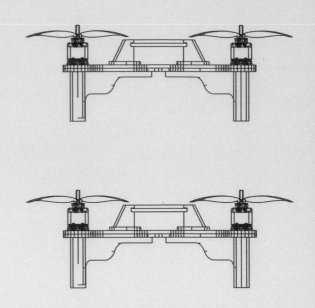

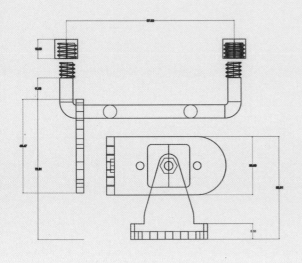

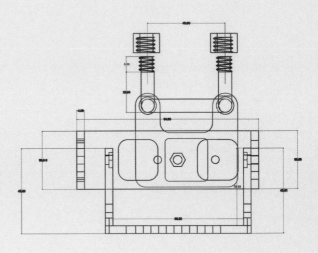

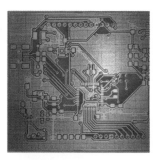

The main board of the Fab Drone project is one of the most complex electronics project ever realized in a Fab Lab, and it required both digital and hand processes.

The frame of the Fab Drone project is a laser cut structure made from plywood.

The Fab Drone requires many cables for the connections among all the components and the main board.

Complex projects like this have a lengthy manufacturing time, in terms of both fabrication and hand assembly.

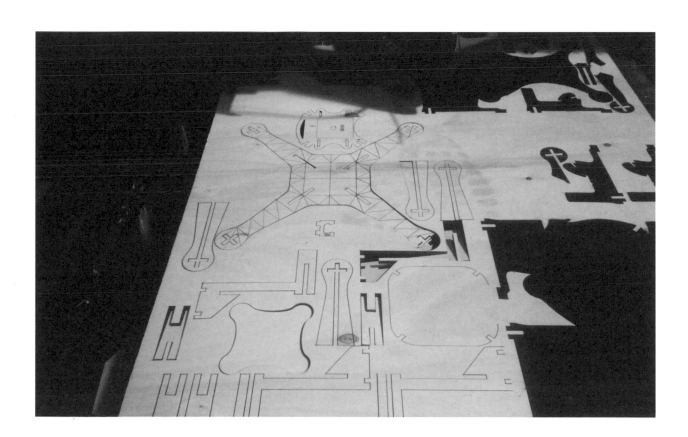

AUTHOR.

Alex Schaub
Angelo Chiacchio
Robert Nelk
Pepijn Fens

DATE.
2011

DESIGNED/FABRICATED.
FABLAB AMSTERDAM
(AMSTERDAM, THE NETHERLANDS)

TECHNOLOGIES USED.
BIG CNC WOOD ROUTER
CNC MILLING
ELECTRONICS PRODUCTIONS
MICROCONTROLLERS
MOLDING AND CASTING
LASER CUTTING
VINYL CUTTING

LINK.
HTTPS://WEB.ARCHIVE.ORG/
WEB/20120614210100/HTTP://FABLAB.WAAG.
ORG/PROJECT/FAB-FOOS
HTTPS://VIMEO.COM/24263449

PROJECT.
Fab Foos

#ENTERTAINMENT

Some sample projects are executed with a view to expanding and communicating knowledge of the technologies and processes available in Fab Labs. One of the best examples of this type of project is Fab Foos, an open source table soccer (also called foosball) table, designed and fabricated in Fablab Amsterdam. This project was made using all the technologies and processes available in a typical Fab Lab, with the exception of 3D printing; it is thus one of the most complex projects to have been developed in a Fab Lab.

The frame of the soccer table is made from recycled wood from the interior of the Fablab Amsterdam building: it was milled with the big CNC wood router on both sides. The players were 3D modeled, their heads based on Michelangelo's David statue. A countermold of two sides was CNC milled from two machinable wax blocks, then the mold was created by casting urethane in the countermold, and the poles inserted in the mold in order to create the hole for it in the players. Each player

was cast with a fast curing resin, colored with red pigments.

The project was intended to be a fully automatic foosball table, with automatic score detection and an interactive scoreboard. To this end, two electronic systems were developed, one for sensing each goal scored, and one to display the scoring system. The first system used an Arduino board and an infrared (IR) receiver/sender couple; these are cheap and small, and can detect when an object – in this case, the ball – passes between the sender and the receiver. The system was designed to be integrated beneath the playing field, making it impossible for players to cheat. The second board consists of a custom CNC milled board and an adhesive circuit cut from a roll of copper sticker with the vinyl cutter. The board controls a series of LEDs inserted just under the surface of the wooden frame, CNC milled to have a very thin surface just above the LEDs, that can then light it from below. The laser cutter was used for few details.

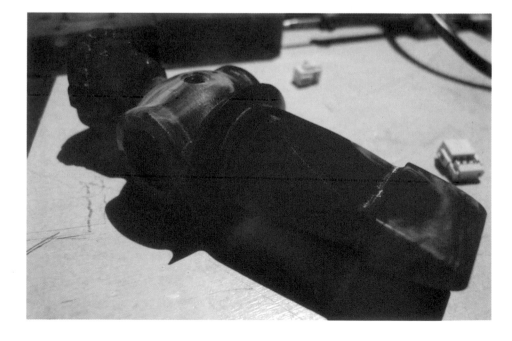

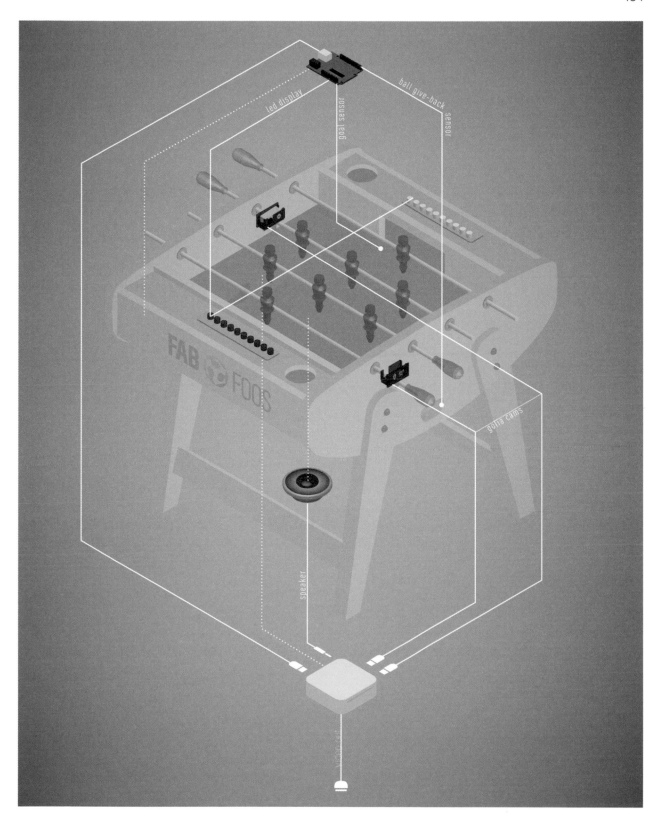

led display

goal sensor

ball give-back sensor

gotta cams

speaker

video out

FAB FOOS

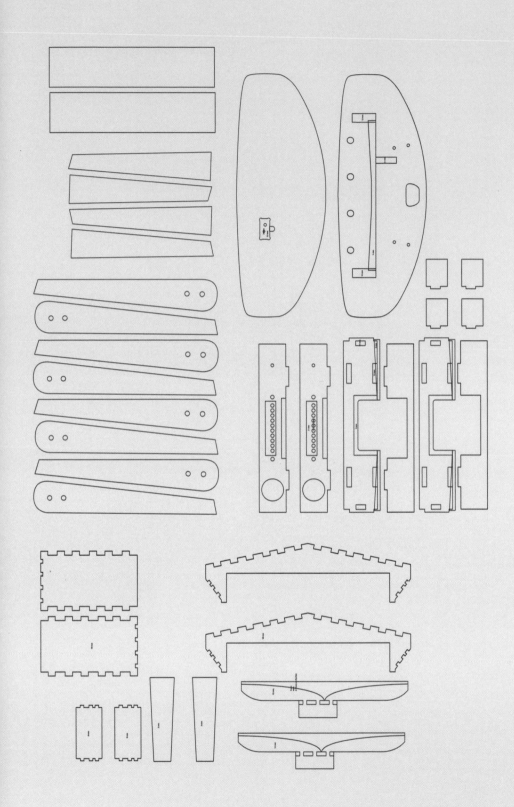

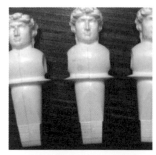

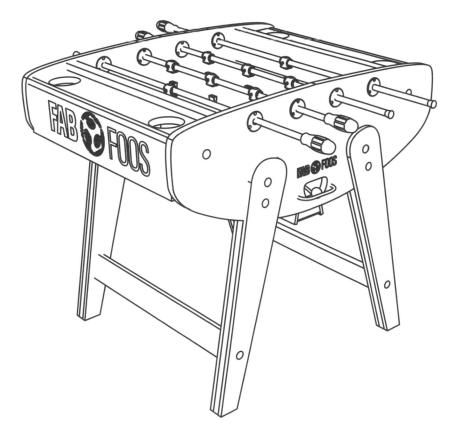

The players in the Fab Foos were modeled on Michelangelo's David statue. This is one of the best examples of how to use molding and casting processes to create multiple copies of an object.

The Fab Foos is now one of the machines available in Fablab Amsterdam, both for manufacturing and for playing together.

The project uses both industrially manufactured Arduino boards and custom digitally fabricated boards.

A 3D model of the final project, fully assembled. Complex projects like these require careful planning from the outset of the design process.

AUTHOR.

IAAC
CBA – MIT
Fab Lab network

DATE.
2010

DESIGNED/FABRICATED.
STRUCTURAL WOOD COMPONENTS MILLED AT
FINNFOREST MERK GMBH (AICHACH, GERMANY)
DESIGNED, COMPONENTS FABRICATED AND
FULLY ASSEMBLED AT FAB LAB BARCELONA
(BARCELONA, SPAIN)
EXHIBITED AT SOLAR DECATHLON EUROPE
(MADRID, SPAIN)

TECHNOLOGIES USED.
BIG CNC WOOD ROUTER
LASER CUTTING
ELECTRONICS PRODUCTION

LINK.
HTTP://WWW.FABLABHOUSE.COM/EN

NOTES.
FULL TEAM CREDITS:
HTTP://WWW.FABLABHOUSE.COM/TEAM-2/

PROJECT.

Fab Lab House

#ARCHITECTURE #ENERGY

While the majority of the projects developed in Fab Labs are quite small, with enough time, resources and collaborators we could even build real houses, amplifying the impact of the Fab Lab on its locality. The proof of this possibility comes from the realization of the Fab Lab House, a 1:1 scale building produced to take part in the Solar Decathlon Europe 2010 competition. Developed by a consortium of organizations and companies led by the Institute for Advanced Architecture of Catalonia (IAAC), The Center for Bits and Atoms at MIT and the Global Fab Lab Network, it is the result of teamwork from people from 25 different countries. The project was designed and fabricated at IAAC and in the Fab Lab it hosts, Fab Lab Barcelona; structural parts were milled in Germany in a wood factory. The house was created with big CNC wood routers and laser cutters; industrial electronics and solar panels were integrated and the building assembled in Fab Lab

Barcelona. It was finally exhibited at Solar Decathlon Europe 2010 in Madrid.
A real solar house, it is made of a 'solar' material, wood, and equipped with solar panels on the roof. It was designed according to the local sunlight patterns: in this project, form follows energy and not aesthetics or function. It is able to produce more than twice the energy it needs, as well as food (in its permaculture garden and fruit trees) and physical goods (in its own small Fab Lab). Sun, water and wind from the environment are managed in order to create a microclimate that passively optimizes the quality of life in the building. It is therefore a self-sufficient building that is active and not passive in generating its own resources. It also has its own kitchen, storage and recycling facilities. Furthermore, the building was designed with a real-time control system that monitors the building's status and interaction with the environment, managing automatically the use and production of energy.

A 3D reconstruction of the final project. This is probably the biggest and most complex project ever realized in a Fab Lab.

Analysis of the sun's rays in the specific locality the house was designed for, both for solar energy and natural lighting.

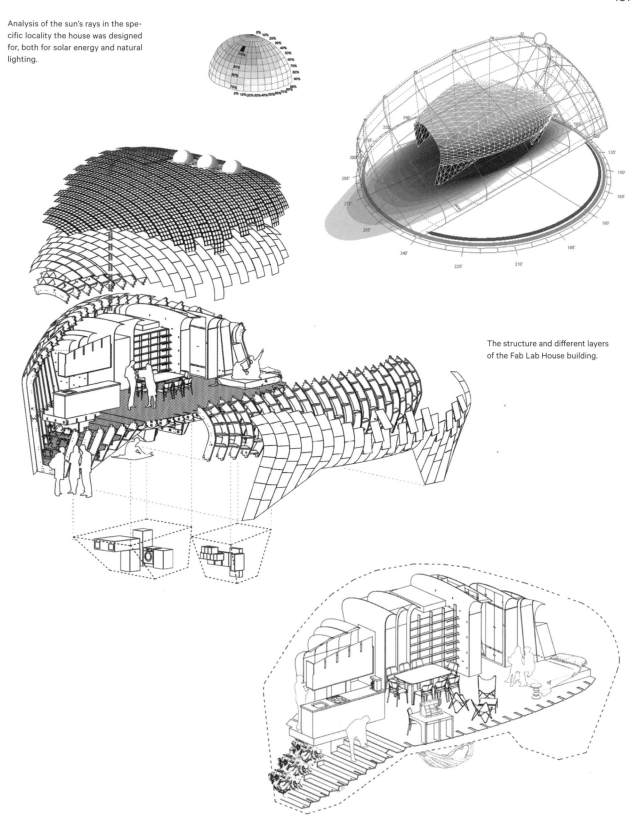

The structure and different layers of the Fab Lab House building.

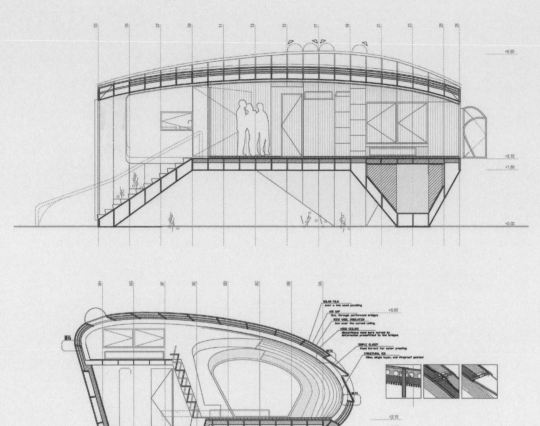

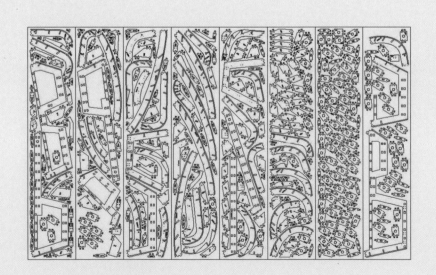

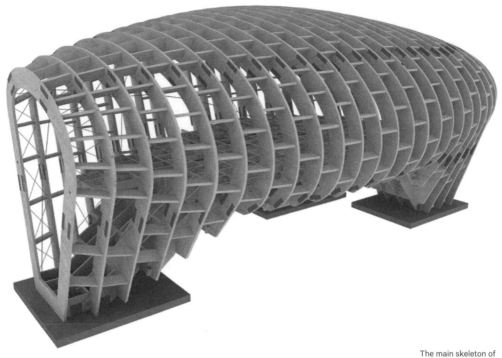

The main skeleton of
the Fab Lab House.

The scale of such a huge project required a lot of processes and coordina-
tion. This project is thus an exception
in the range of Fab Lab projects, but
still something that can be achieved
within Fab Labs, especially when they
collaborate with external manufacturing facilities.

AUTHOR.

Luciano Betoldi

PROJECT.
Fab Skate

#SPORT #ENTERTAINMENT
#TRANSPORTATION

DATE.
2013

DESIGNED/FABRICATED.
FAB LAB BARCELONA (BARCELONA, SPAIN)

TECHNOLOGIES USED.
BIG CNC WOOD ROUTER
THERMOFORMING
LASER CUTTING

LINK.
HTTPS://WEB.ARCHIVE.ORG/
WEB/20131031234912/HTTP://WWW.FAB-
SKATE.COM/SITE/FABSKATE.HTML

HTTPS://WWW.YOUTUBE.COM/
WATCH?V=3C4SJOTU5WU

PHOTOS.
LUCIANO BETOLDI

A process that has been emerging recently on the Fab Lab scene involves milling a big mold with a big CNC wood router, in combination with a thermoforming machine to create flat or curved surfaces. A very good example of this process is the Fab Skate project, a series of workshops with young girls and boys that has been organized at Fab Lab Barcelona. During these workshops, the kids learn how to build fully customized, high-quality skateboards and longboards for the price of an off-the-shelf model. This customization is derived from the size, the curvature and the decoration or engraving of the surface.

The big CNC wood router can be used to shape a two-sided polyurethane mold. This is used inside the thermoforming machine, where three 4 mm bamboo wood layers are mixed with epoxy resins from pine trees, gluing them and fixing the curve. After a while, the epoxy resins dry and the board maintains its curve. Finally, the laser cutter can be employed to engrave custom drawings on the board. These projects and workshops are also indicative of the ever-increasing tendency to enfranchise younger people and children within the Fab Lab movement.

The final outcome of the whole process: it is a real skateboard, and not just a prototype.

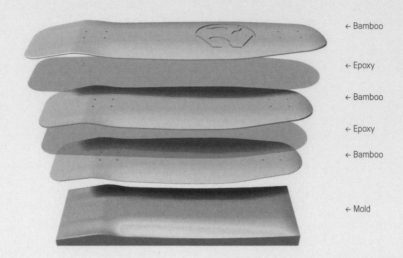

← Bamboo

← Epoxy

← Bamboo

← Epoxy

← Bamboo

← Mold

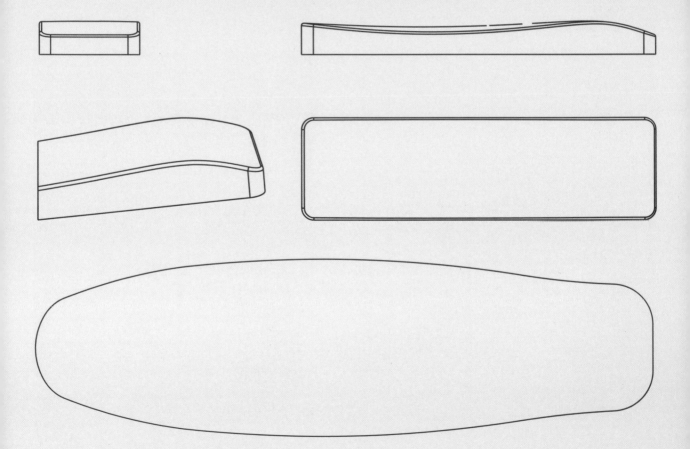

After the development of this project, many Fab Labs tried to build their own version of a skate. But to date this is still the most complex and multi-layered.

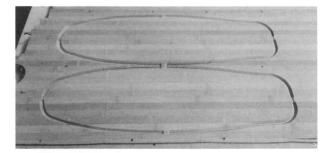

The surface finishing of the Fab Skate shows how carefully it was fabricated, and how it is indistinguishable from an industrial one.

AUTHOR.

Jens Dyvik

PROJECT.
Layer Chair

#FURNITURE

DATE.
2012

DESIGNED/FABRICATED.
DESIGNED AT HONFABLAB (FABLAB YOGYA-
KARTA) (YOGYAKARTA, INDONESIA);
CUSTOMIZED AND FABRICATED AT FABLAB AM-
STERDAM (AMSTERDAM, THE NETHERLANDS),
FABLAB LYNGEN (ØRNES, LYNGSEIDET, NORWAY),
FABLAB SEVILLA (SEVILLE, SPAIN), FAB LAB
WELLINGTON (WELLINGTON, NEW ZEALAND)

TECHNOLOGIES USED.
BIG CNC WOOD ROUTER

LINK.
HTTP://WWW.DYVIKDESIGN.COM/SITE/
PORTFOLIO-JENS/PRODUCTS/THE-LAYER-CHAIR
HTTP://WWW.DYVIKDESIGN.COM/SITE/
PORTFOLIO-JENS/THE-LAYER-CHAIR-
AMSTERDAM-EDITION.HTML
HTTPS://WWW.YOUTUBE.COM/
WATCH?V=PNR1YBIGQCY

PHOTOS.
JENS DYVIK

Fab Labs are part of a global network, within which each node collaborates with the others. But this collaboration doesn't arise of its own accord: it's up to each Fab Lab to be active in the network. Most of the time, the network among the nodes is built by dynamic individuals who travel among Fab Labs, creating collaborative projects that strengthen these connections. A great example of such 'Fab Nomads' is Jens Dyvik, a Norwegian designer who has also studied and worked in the Netherlands. Jens traveled among many countries and Fab Labs, creating and consolidating connections among them. He documented his experience in *Making Living Sharing,* a Fab Lab world tour documentary, which can be seen on YouTube.

Several projects were collaboratively created, developed and shared during these travels. Perhaps the most representative and relevant project is the Layer Chair, which has become one of the most replicated projects in the Fab Lab network. While working at HONFablab in Indonesia, Jens Dyvik created a parametric chair. Different settings – such as the sitting profile, width, height, material thickness and so on – can be adjusted before fabrication. The chair is made up of several different layers or 2D sections that are glued together with the help of holes and alignment pins in order to form a 3D object. The layers are CNC milled out of a big wood or MDF sheet with a big CNC wood router. Jens Dyvik then worked for a while at Fablab Amsterdam in the Netherlands, and there he developed the project further and fabricated

The basic structure of the Layer Chair, that was modified and fabricated in a different way in many Fab Labs.

The Layer Chair can be fabricated in
different materials, colors and shapes.

The Layer Chair, as fabricated in
FabLab Lyngen in Northern Norway.

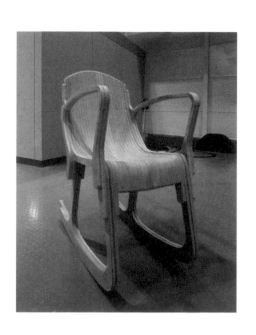

The Layer Chair as manufactured
in Fablab Amsterdam.

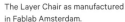

different versions, starting with 18 mm thick black MDF as the main material. There he adapted the design to build a similar table and a custom chair for the world famous cellist Frances Marie Uitti, which she could use for playing the cello. Then Dyvik traveled to FabLab Lyngen in northern Norway, where he created 15 chairs with 27 mm thick pinewood plates. Each chair has a different profile, each one taking inspiration from the surrounding mountains. He then brought the project to Fablab Sevilla in Spain, where three groups of students downloaded and customized their own versions of the Layer Chair. He created a digital system that 3D scans people's bodies and creates a custom design based on the data. The Layer Chair was also transformed into the Layer Stool by another student at Fab Lab Wellington in New Zealand.

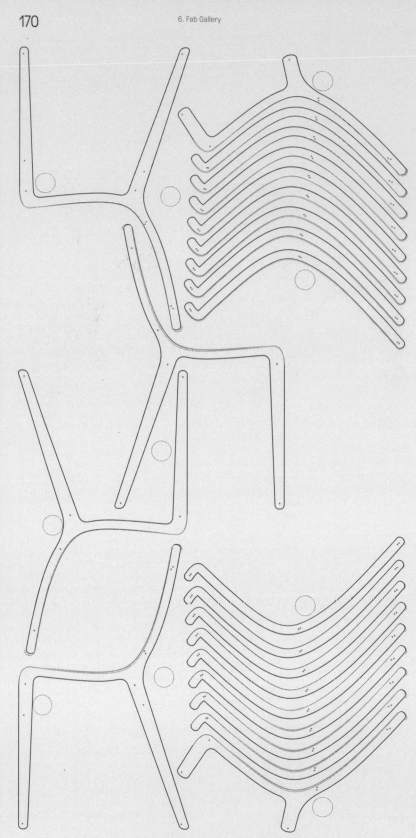

The complexity of the Layer Chair can be seen in the parametric setup in Rhinoceros / Grasshopper, which can be modified in order to generate many different designs. The whole idea is simple and cheap: the chair can be milled out of cheap plywood or MDF and then all the layers glued together.

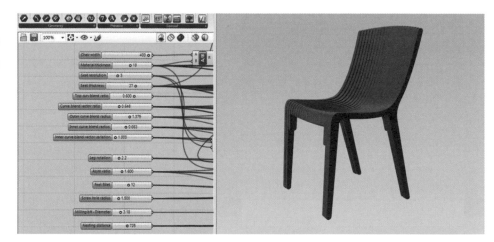

AUTHOR.

Pietro Leoni

PROJECT.

Lettera

#MACHINE #ART #EDUCATION

DATE.
2013

DESIGNED/FABRICATED.
FABLAB TORINO (TURIN, ITALY)

TECHNOLOGIES USED.
LASER CUTTING
ELECTRONICS PRODUCTION
MICROCONTROLLERS

LINK.
HTTP://PIETROLEONI.COM/#LETTERA

PHOTOS.
PIETRO LEONI

One of the most interesting aspects of CNC machines is that the same concept of a computer numerically controlled machine can be applied to many different end tools and media. There is a lot of experimentation going on in Fab Labs around this topic, involving both the hacking of existing machines and the design of new custom machines. One of the most common experiments is the transformation of the CNC machine concept into a drawing machine, fusing the digital side of computer control with the analog side of pencils, pens, brushes and the like. Most of the time these experimentations are concerned only with the technical side of the project, but sometimes designers also work on the aesthetic and semiotic sides, conceiving new machines that not only create artistic objects but that are also linked to existing artistic objects. One of these rare examples is Pietro Leoni's

Lettera project, designed and fabricated at Fablab Torino, in Turin, Italy. The shape and the name of the project are a reminder of the Lettera 22, an excellent example of Italian product design that was produced not far from Turin several decades ago. The Lettera 22 is a portable mechanical typewriter designed by Marcello Nizzoli in 1950 and produced by Olivetti. It won the most prestigious design prize in Italy, the Compasso d'oro, in 1954. The Lettera project is a portable CNC drawing machine that can control any kind of pen, pencil or brush: it is made of a laser cut MDF sheet and is controlled by an Arduino. It is composed of the highest possible number of laser cut components, on a budget of less than €100. This is not only a great example of a CNC drawing machine, but also a simple system to assist students in learning the principles of numerical control machines and digital production.

A 3D reconstruction of Lettera shows us where the pen is inserted and the working area for drawing custom shapes on paper.

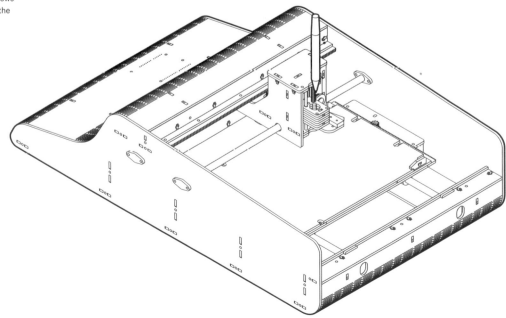

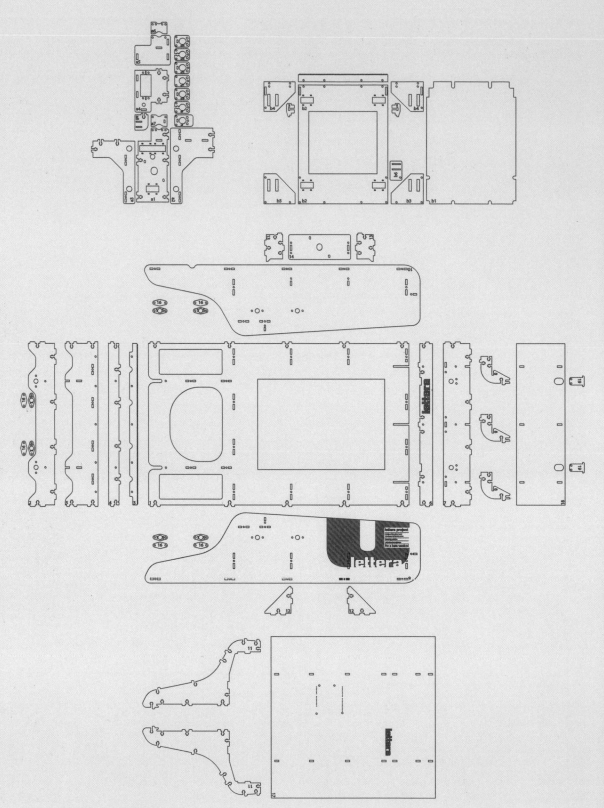

Completely laser cut out of MDF, the Lettera is quite cheap and easy to make. The complexity of the project lies in the many components that are needed to build it.

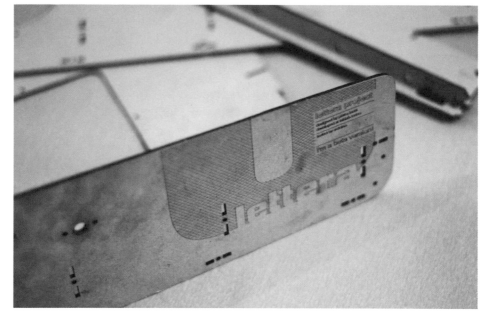

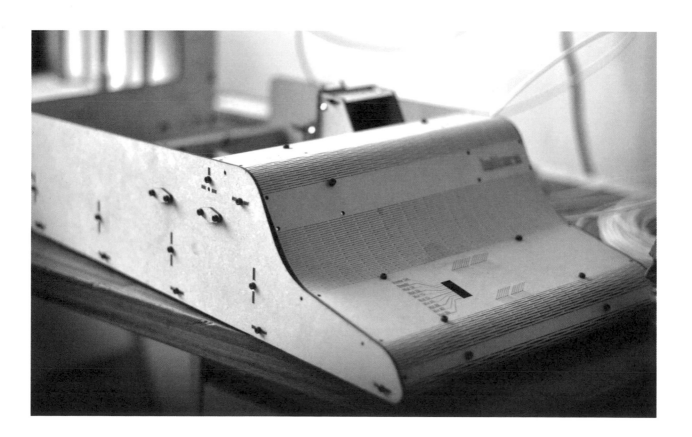

The fully assembled Lettera, without the Arduino board that controls its movements.

AUTHOR.

Yradia

PROJECT.

#DATA #NATURE #FOOD #MACHINE

MEG - Micro Experimental Growing

DATE.
2012

DESIGNED/FABRICATED.
YRADIA (MILAN, ITALY)
WITH THE COLLABORATION OF
MAKEINBO (BOLOGNA, ITALY)

TECHNOLOGIES USED.
ELECTRONICS PRODUCTION
MICROCONTROLLERS

LINK.
HTTP://WWW.GROWMEG.ORG/
HTTPS://WWW.FACEBOOK.COM/MEG.AGTECH/
HTTPS://WWW.KICKSTARTER.COM/PROJECTS/
YRADIA/MEG-OPEN-SOURCE-INDOOR-
GREENHOUSE/

TEAM.
PIERO SANTORO, CARLO D'ALESIO,
ANDREA SARTORI, MASSIMO DI FILIPPO,
FEDERICO CASOTTO

PHOTOS.
MASSIMO DI FILIPPO, JESSICA MAULLU

Sometimes even professional design consultancy firms develop projects using open source, open hardware and open design philosophies, often with the help of Fab Labs, or on the principle that such projects could also be manufactured in Fab Labs. Micro Experimental Growing (MEG) is one of these cases: it constitutes an open source, fully automatized indoor greenhouse. Developed by Yradia, an innovation consultancy firm dedicated to lighting, MEG is a complex project that involves photometrics, mechanical and industrial design, electrical design, PCB design, and GUI+Ux design.

This project started as an independent research project on lighting spectrum optimization to maximize Photosynthetically Active Radiation (PAR) for indoor growing. MEG is an open source machine that runs on Arduino and can be controlled with an iPad, iPhone or Android phone or tablet.

Users use these devices to communicate with many sensors and actuators that manage light cycle, air circulation, temperature and heat generation and recycling, humidification, nutrient distribution, watering, and soil pH sensing. Each MEG can also be wirelessly interconnected with other MEGs. MEG's Light Engine was designed from scratch as a fusion between expensive, state-of-the-art technology and easily accessible and democratic technology.

MEG allows users to control precisely the key parameters for growing plants. These parameters can be shared with other users and researchers via a dedicated online platform, thus contributing to the development of a shared knowledge database accessible to any MEG user. This device could be applied in harsh environments, but also for research, education, space farming, the growing of medical plants, and in other fields.

MEG can adapt and change its lighting system by combining different colors and intensities. This is both a functional and aesthetic effect, defining as it does the visual impact of the MEG on its environment.

OVERALL DIMENSIONS AND TECHNICAL DETAILS

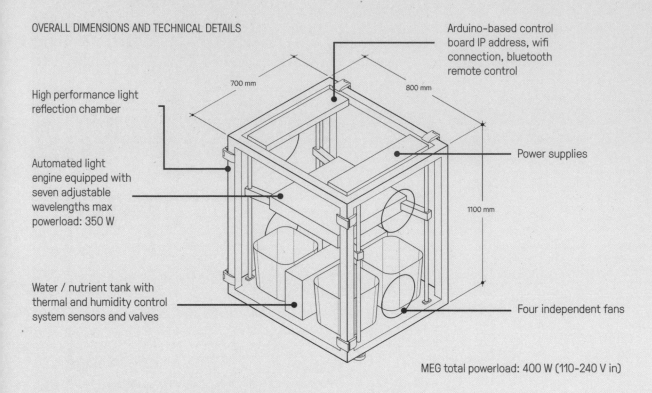

Arduino-based control
board IP address, wifi
connection, bluetooth
remote control

High performance light
reflection chamber

700 mm

800 mm

Power supplies

Automated light
engine equipped with
seven adjustable
wavelengths max
powerload: 350 W

1100 mm

Water / nutrient tank with
thermal and humidity control
system sensors and valves

Four independent fans

MEG total powerload: 400 W (110-240 V in)

FUNCTIONS - PARAMETERS THAT CAN BE SET THROUGH THE GUI

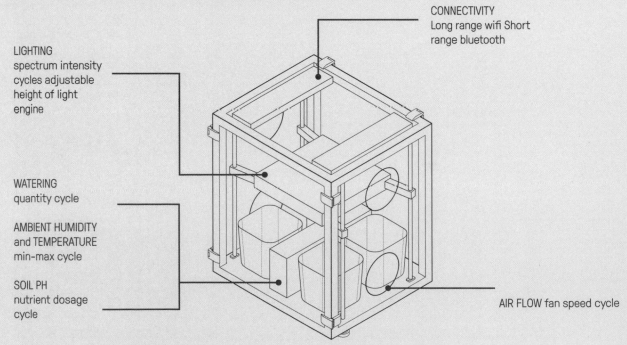

CONNECTIVITY
Long range wifi Short
range bluetooth

LIGHTING
spectrum intensity
cycles adjustable
height of light
engine

WATERING
quantity cycle

AMBIENT HUMIDITY
and TEMPERATURE
min-max cycle

SOIL PH
nutrient dosage
cycle

AIR FLOW fan speed cycle

DOWNLOAD. http://www.growmeg.org/build-your-own

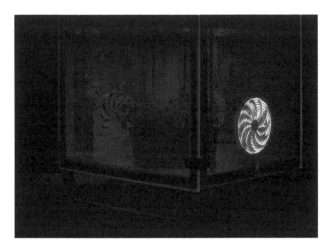

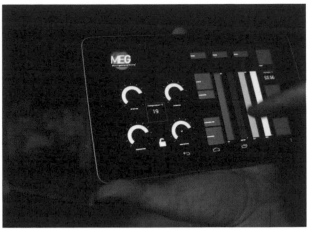

MEG consists of industrially-produced components and advanced lighting technologies and functionalities. This is a great example of how open design and open hardware projects can range from simple DIY to very complex projects developed by highly-trained professionals.

Future versions will be developed with the collaboration of universities and other Fab Labs like Opendoc. As an example, a custom version was developed with them for the Expo 2015 in Milan, Italy, where five machines were installed in the subway. This further demonstrates the potential of the project to make agriculture possible, even in harsh environments.

Furthermore, the project was carefully designed to be an aesthetically pleasing object, not just a technical device.

AUTHOR.

Stefano Maffei
Tecnificio
Various designers*

DATE.
2014

DESIGNED/FABRICATED.
DIGITAL FABRICATION IN THE FAB LAB
MAKE IN MILANO (MILAN, ITALY)
CRAFT MANUFACTURING WITHIN THE SLOW/D
NETWORK (ITALY)
EXHIBITION IN SUBALTERNO1 (MILAN, ITALY)

TECHNOLOGIES USED.
PRODUCT HACKING
3D PRINTING: FUSED DEPOSITION MODELING
(FDM)
HAND MANUFACTURING

LINK.
HTTP://WWW.SUBALTERNO1.COM/

NOTES.
CURATOR: STEFANO MAFFEI
ART DIRECTION: TECNIFICIO
*DESIGN: MASSIMILIANO ADAMI + FRANCESCO
TARANTINO, FRANCESCO BOMBARDI + MAR-
CELLO LIGABUE, FRANCESCA LANZAVECCHIA
+ HUNN WAI, CLAUDIO LARCHER, SIMONE SIM-
ONELLI + STEFANO CITI, BRIAN SIRONI + GIULIA
TACCHINI, ALESANDRO STABILE, TECNIFICIO

PHOTOS.
SUBALTERNO1

PROJECT.
MONDOPASTA

#FOOD #MACHINE #TOOL

Fab Labs sometimes collaborate with traditional artisans, mixing digital fabrication and handmade processes. MONDO-
PASTA is a project initiated for the Salone
del Mobile 2014 (the Milan Furniture Fair).
This was an an exhibition curated by the
Subalterno1 gallery (dedicated to Italian
self-production designers, in other words,
designers that produce and commercial-
ize their products independently) with
the collaboration of the Fab Lab Make in
Milano (one of the Fab Labs in Milan), Tecnificio (not a Fab Lab but a Maker facility)
and Slow/D (a platform that brings together designers and artisans in distributed
manufacturing networks).
The exhibition invited seven emerging
designers from Italy (including Fab Lab
Reggio Emilia and Tecnificio) to develop

experimental and innovative projects that
would redefine the geometry, process
and meaning of pasta, one of Italy's most
emblematic products. The projects mixed
3D printing in FDM with product hacking
of existing pasta machines or 3D printing
machines to create new tools for working
with pasta.
Projects included a machine for creating spaghetti with different thicknesses
for different cooking times, a converted
3D printer with a knife for CNC pasta creation, a machine for creating 3D
pasta made of a single strand of spaghetti, wooden robots for working with
pasta, stamps for custom pasta, a set of
3D printed cutting wheels with different
designs, and a 3D printed tattoo gun that
can decorate pasta with squid ink.

Projects like MONDOPASTA show how
even a small Fab Lab can be a creative
place for regenerating old and established traditions.

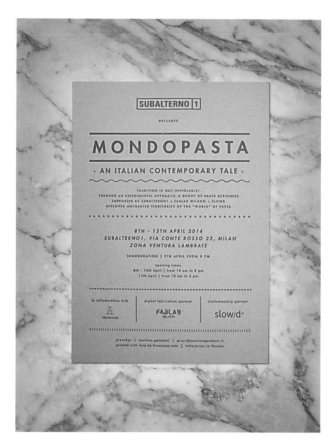

MONDOPASTA started as an initiative of the Subalterno1 gallery, and participated in the European Maker Faire in Rome in 2014. Custom projects were developed and fabricated in Fab Labs and other workshop, and the process also included working in a real kitchen, to test new proposals for pasta recipes.

The Bipasta project (Brian Sironi - Giulia Tacchini) consists of a new recipe for pasta, one which includes crickets as a sustainable food source. The designers created a 3D printed mold to introduce this new recipe to users.

The functionalities of existing tools can be applied to different materials. The Nigredo project (Patrizia Bolzan - Marcello Pirovano) is based on a customized professional tattoo machine that can be used for creating drawings not on human skin but on pasta.

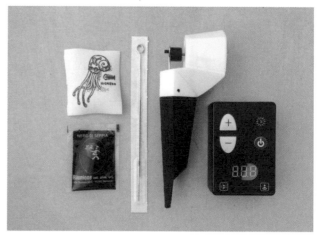

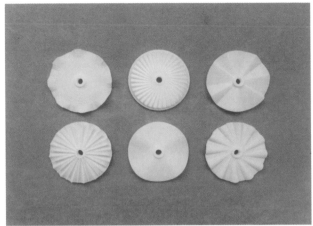

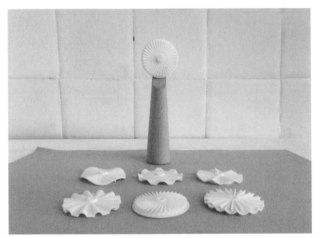

Using even cheap 3D printing FDM machines it is possible to modify the geometry of traditional tools. And with new tools, new kinds of pasta can be created, with different flavors and qualities. The Stracottoaldente (Massimiliano Adami - Francesco Tarantino) and Zag (Alessandro Stabile) projects are an example of this approach.

A common FDM 3D printer can be hacked and modified in order to work as a pasta cutting tool, but with computer numerically controlled precision. The Nonnabot (Francesco Bombardi - Marcello Ligabue) represents a new kind of digital fabrication machine.

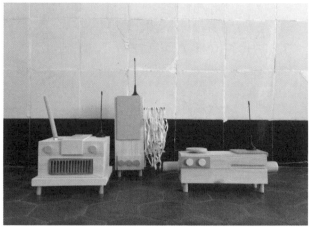

Sometimes digital fabrication machines can be hacked and reused as analog machines, without any digital control, and still be innovative. This is the case of the Uncut (Francesca Lanzavecchia - Hunn Wai) project.

The projects developed in Fab Labs don't have to be exclusively digital and digitally fabricated. The Dr. K collection (Claudio Larcher) is an example of this: it is a collection of fake kitchen machines that contain simple and universal tools to make pasta.

AUTHOR.

Teja Philipp
Jennifer Heier
Philipp Engel

PROJECT.
Mr Beam

#MACHINE

DATE.
2014

DESIGNED/FABRICATED.
MUNICH, GERMANY

TECHNOLOGIES USED.
ELECTRONICS PRODUCTION
MICROCONTROLLERS
3D PRINTING: FUSED DEPOSITION MODELING
(FDM)
LASER CUTTING

LINK.
HTTP://MR-BEAM.ORG/
HTTPS://WWW.KICKSTARTER.COM/PROJECTS/
MRBEAM/MR-BEAM-A-PORTABLE-LASER-
CUTTER-AND-ENGRAVER-KIT

Many projects developed within or for Fab Labs are digital fabrication machines themselves, in a constant evolution of more and more accessible technologies for production. Many projects are based around CNC milling machines, big CNC wood routers and FDM 3D printers, but there are some projects for other technologies. The Mr Beam laser cutter is a very interesting experiment in this field, with a highly successful crowdfunding campaign on Kickstarter, from which almost $180,000 was raised. The goal of the project is to maximize the working area while keeping costs to a minimum (two sizes are available: ANSI C / DIN A2 or letter size), while also making the machine portable, easy to build and simple to use. Inspired by other open source laser cutter projects and the RepRap project, Mr Beam is sold as a kit that can be assembled with standard tools; no soldering is required.

Mr Beam is designed in a way that is distinct from traditional CO_2 laser cutters: it can be moved around directly above the workpieces, there are no optics that reflect the laser beam, and the height-adjustable legs allow the processing of materials of variable thickness, sometimes in multiple passes. Mr Beam is operated via a web interface that runs on a Raspberry Pi, while the stepper motors are controlled by an Arduino Uno equipped with a custom board. Using a standard network cable, the device can be connected to a home router, accessed on a web browser with a computer, tablet or smartphone, or directly connected to a computer.

Mr Beam could have a huge impact on the teaching of digital fabrication, since the mechanism and the process are much more visible here than in industrial machines. Even the laser beam can be clearly seen.

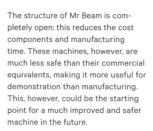
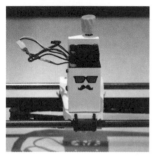

The structure of Mr Beam is completely open: this reduces the cost components and manufacturing time. These machines, however, are much less safe than their commercial equivalents, making it more useful for demonstration than manufacturing. This, however, could be the starting point for a much improved and safer machine in the future.

AUTHOR.

Joel Gibbard

PROJECT.

Open Hand Project

#HEALTH #ROBOT

DATE.
2014

DESIGNED/FABRICATED.
BRISTOL (UNITED KINGDOM)

TECHNOLOGIES USED.
3D PRINTING: FUSED DEPOSITION MODELING
(FDM)
ELECTRONICS PRODUCTION
MICROCONTROLLERS

LINK.
HTTP://WWW.OPENHANDPROJECT.ORG/
HTTP://WWW.INSTRUCTABLES.COM/ID/
DEXTRUS-V11-ROBOTIC-HAND/
HTTP://WWW.THINGIVERSE.COM/THING:287638

Many Fab Labs are working on the development of prostheses, especially open source, low-cost and customizable ones, which could be manufactured and distributed globally at a fraction of the cost of existing commercial (often prohibitively expensive) versions. For example, many robotic and prosthetic hands can be easily created with the technologies available in Fab Labs. One of the most advanced and best documented of such initiatives is the Open Hand project, the goal of which is to make robotic prosthetic hands more accessible to amputees. Leading prosthesis can cost up to $100,000, but by using FDM 3D printers, and producing the objects using strong ABS plastic, the price can be reduced to under $1000. The project is completely open source, and any company will therefore be able to use the designs and sell the hands all over the world.

The main model in the Open Hand project is the Dextrus hand. It uses electric motors in place of muscles, and steel cables in place of tendons: these parts are controlled by electronics, giving it a natural movement that can handle all sorts of different objects. The hand can be connected to an existing prosthesis using a standard connector. The main difference between the Dextrus hand and many other 3D printed and low-cost prostheses is that most of the time they are purely mechanical devices, with no motors or electronics. Dextrus, by contrast, features motors and electronics, with a battery able to operate for many hours. It can also be connected to a computer through a standard USB connection and used for many purposes: animatronics, education, research. Each finger can be articulated individually with a feedback sensor, enabling it to grasp all sorts of different shapes and sizes.

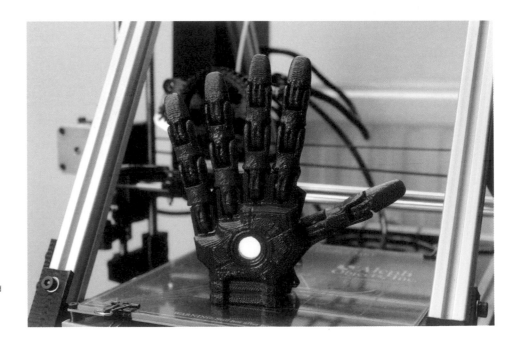

An example of one of the Open Hand project models on top of a FDM 3D printer used to manufacture its components.

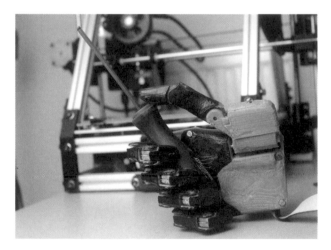

A closer look at the manufactured Dextrus model reveals that the object was created with an FDM 3D printer. The imperfections on the surface are purely aesthetic; objects printed using this technology are both cheap and of a high standard in terms of mechanics.

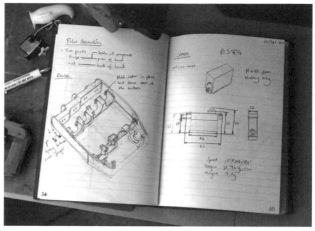

AUTHOR.

Fab Academy version
OpenSourceBeehives.net

PROJECT.
Open Source Beehives

#DATA #NATURE #FOOD

DATE.
2013

DESIGNED/FABRICATED.
FAB LAB BARCELONA (BARCELONA, SPAIN)
GREEN FAB LAB (VALLDAURA – BARCELONA, SPAIN)
OPEN TECH COLLABORATIVE (DENVER, USA)

TECHNOLOGIES USED.
BIG CNC WOOD ROUTER
LASER CUTTING
CNC MILLING
ELECTRONICS PRODUCTION
MICROCONTROLLERS

LINK.
HTTP://WWW.OPENSOURCEBEEHIVES.NET/
HTTP://FABACADEMY.ORG/ARCHIVES/2013/
STUDENTS/REES.JOHN/INDEX.HTML

NOTES.
FAB ACADEMY VERSION: JOHN REES,
JONATHAN MINCHIN, FERRAN MASIP VALLS
OPENSOURCEBEEHIVES.NET:
JONATHAN MINCHIN, TRISTAN COPLEY-SMITH,
AARON MAKARUK, SENECA KRISTJONSDOTTIR,
GARRETT BURRG, CHRIS BORKE,
PATRICK BESEDA

A growing number of projects developed in Fab Labs are focused on questions of sustainability, and one of the most interesting of these is the Open Source Beehives project, which is based on a network of citizen scientists tracking the decline in bee populations. This project started as a collaboration between Fab Lab Barcelona (Barcelona, Spain) and Annemie Maes from OKNO (Brussels, Belgium), and later became a collaboration between Fab Lab Barcelona, Green Fab Lab / Valldaura (Barcelona, Spain) and Open Tech Collaborative (Denver, USA). This collaboration started as a Fab Academy collaborative project (carried out by John Rees, Jonathan Minchin and Ferran Masip) and then evolved into a bigger, stable team led by Jonathan Minchin and partners Tristan Copley Smith and Aaron Makaruk in Denver. It has since become a registered company.

The Open Source Beehives project has developed two open source beehive designs: the Colorado Top Bar and the Barcelona Warré. All the files are based entirely on open source software, hardware and design: anybody can download the information and manufacture and install a beehive locally. The hives can be manufactured with big CNC wood routers, with further components and details created with laser cutting and 3D printing (FDM). The digital part is based on a sensor kit that allows users to monitor bee colony information and upload the data to the online repository called the Smart Citizen platform, where it can be viewed and downloaded by anybody. The hardware embedded in the beehives can detect colony mood (using audio sensors), pesticide (using volatile organic compound sensors), bee numbers (using infrared sensors), hive humidity (using humidity sensors), hive temperature (using temperature sensors) and hive location (using GPS devices). It also employs cameras connected to the Internet to provide visual data about the colony.

After the design and manufacturing of the first models, a test process was launched.

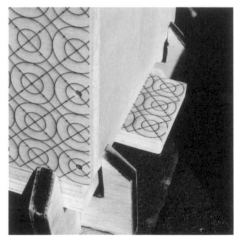

The goal of the test is to understand the impacts of the projects on bee colonies.

The structures were designed using the Open Structure modular grid, laser engraved on the surface.

The Open Source Beehive is not just a prototype: it is already in use in some places.

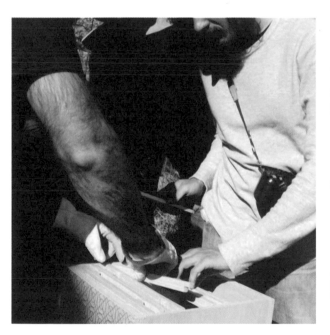

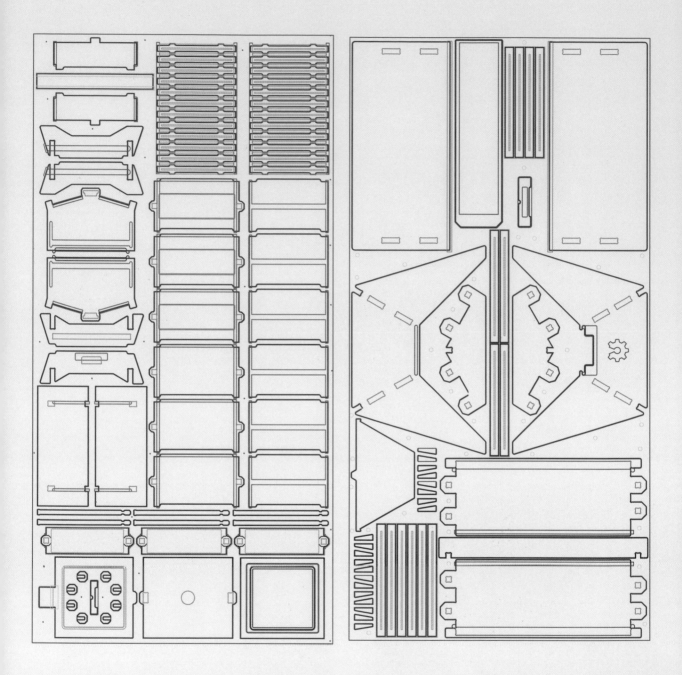

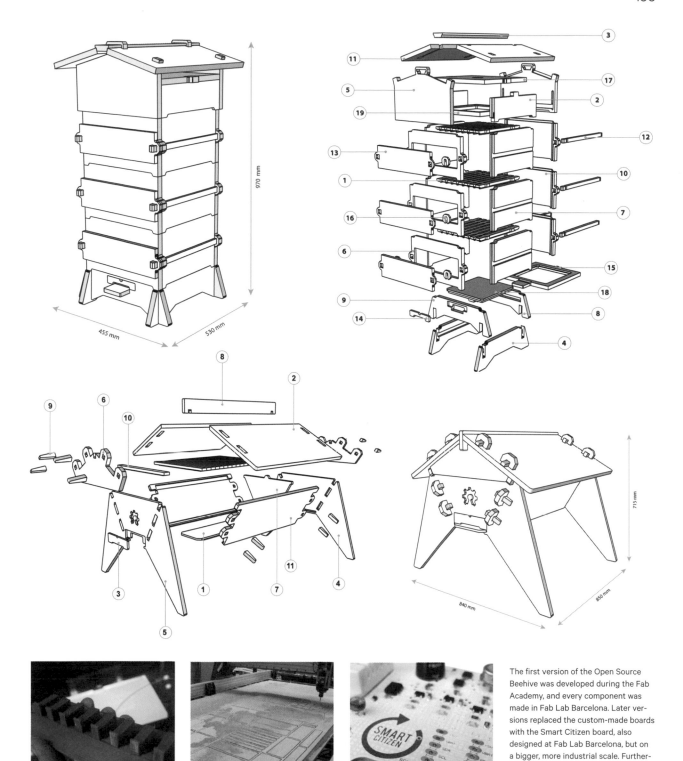

The first version of the Open Source Beehive was developed during the Fab Academy, and every component was made in Fab Lab Barcelona. Later versions replaced the custom-made boards with the Smart Citizen board, also designed at Fab Lab Barcelona, but on a bigger, more industrial scale. Furthermore, it already has its own software and online platform for gathering data.

AUTHOR.

Gerard Rubio Arias

PROJECT.
OpenKnit

#MACHINE #FASHION

DATE.
2014

DESIGNED/FABRICATED.
FAB LAB BARCELONA (BARCELONA, SPAIN)

TECHNOLOGIES USED.
3D PRINTING: FUSED DEPOSITION MODELING
(FDM)
LASER CUTTING
ELECTRONICS PRODUCTION
MICROCONTROLLERS

LINK.
HTTP://OPENKNIT.ORG/
HTTPS://GITHUB.COM/G3RARD/OPENKNIT

OpenKnit is an open source software, hardware and design digital fabrication tool, which enables the user to create her own bespoke clothing from digital files. A low-cost device (under €550), OpenKnit is a fully automated, accessible knitting machine that can be used to create knitted fabrics. There are very few accessible technologies for knitting in Fab Labs, so projects like this could have a significant impact on how fashion design projects can be undertaken within the Fab Lab network.
The project is based on other open source projects: it uses an Arduino Leonardo board, and the Knitic software, developed by Mar Canet and Varvara Guljajeva (http://www.knitic.com/), which can be used as a CAM for the knitting machine. Furthermore, digital cloth designs can be shared on the online repository DoKnitYourself, created by Takahiro Yamaguchi (http://doknityourself.com/). We are witnessing an emerging technology ecosystem for open source knitting machines; this is an open source project that enables Fab Lab users to develop even more open source projects in the field of fashion design.

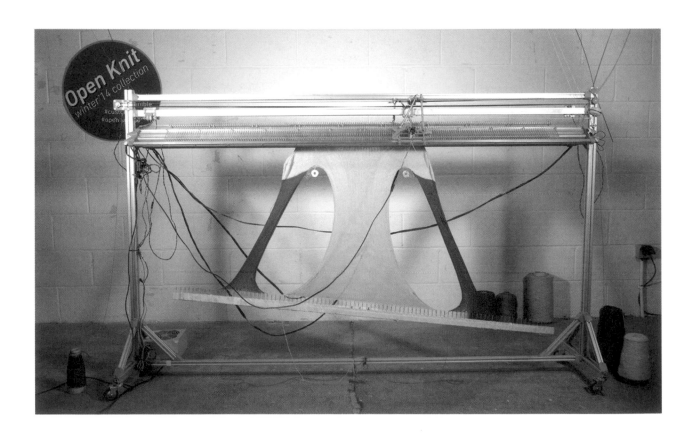

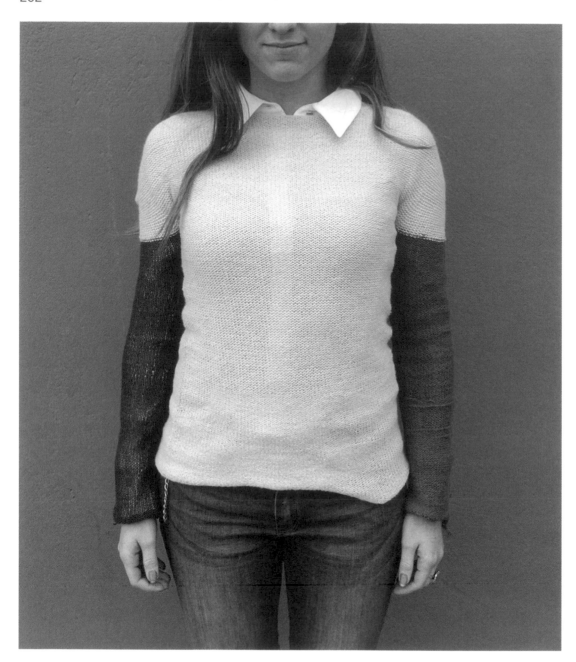

The clothes produced with OpenKnit
are not just prototypes, but finished,
ready-to-wear garments.

The development of the OpenKnit project is a continuous process, based on trying new clothes and involving atypical users.

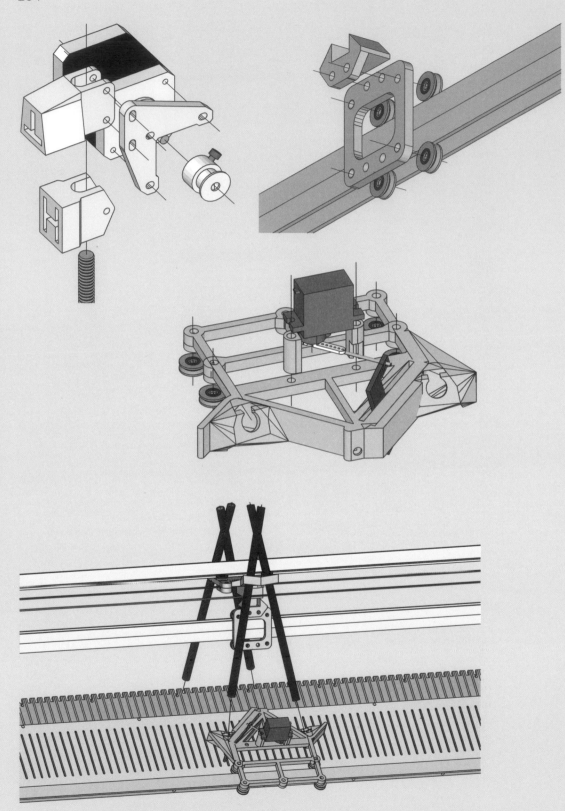

The OpenKnit project consists of few small components that can be laser cut or 3D printed. Its simple design makes it easy to manufacture, which is why it has been adopted by many Fab Labs.

AUTHOR.

Flowers–INRIA

PROJECT.

Poppy Project

#ROBOT #RESEARCH #EDUCATION

DATE.
2013

DESIGNED/FABRICATED.
FLOWERS (FLOWING EPIGENETIC ROBOTS AND
SYSTEMS)
INRIA (TALENCE, FRANCE)
ENSTA PARISTECH (PARIS, FRANCE)

TECHNOLOGIES USED.
ELECTRONICS PRODUCTION
MICROCONTROLLERS
ANY 3D PRINTING TECHNOLOGY

LINK.
HTTP://WWW.POPPY-PROJECT.ORG/

TEAM.
PIERRE YVES OUDEYER, MATTHIEU LAPEYRE,
PIERRE ROUANET, JONATHAN GRIZOU, STEVE
NGUYEN, ALEXANDRE LE FALHER, FABIEN
DEPRAETRE

The Poppy project was designed by the Flowers Lab at INRIA Bordeaux and ENSTA ParisTech (France). It constitutes a low-cost, hackable humanoid robot. It is composed of three principal elements, namely, its mechatronic structure (skeleton and motors), its electronics, and its software. The software, hardware and design that comprise these three elements are all open source. In addition, Poppy was designed with off-the-shelf motors, and electronics and parts that can be fabricated using any 3D printing technology. Assembly takes approximately two days, and costs in the region of €7500. Because the hardware is based on an Arduino board, extra sensors and devices (like cameras and LCD screens) can be integrated without much difficulty, in order to optimize its capacity to register the environment, and interact socially with users. A computer using open source PyPot library (written in Python and available for Linux, Windows and Mac OS) can be used to control Poppy. The design of the body is derived from the human skeleton: bending legs, a multi-articulated trunk and a soft body, allowing for greater stability and agility in motion, and greater robustness of form. Although Poppy wasn't, in fact, designed or fabricated in a Fab Lab, the development of the project was undertaken keeping the Fab Lab network in mind. The electronics needed for this project are rather challenging (too much so for home-production); in addition, INRIA (the French Institute for Research in Computer Science and Automation) only carries out research. The team has therefore been developing a process and business model that relies on Fab Labs for the local production and distribution of the robot. Fab Labs will be able to market fully-assembled and functional Poppy robots, as well as offer support, repair, upgrade and customization services. The project could also give rise to events or artist residencies in Fab Labs.

The humanoid of the Poppy project is a small scale robot, which lowers manufacturing costs and makes it easy to assemble and to experiment and work with.

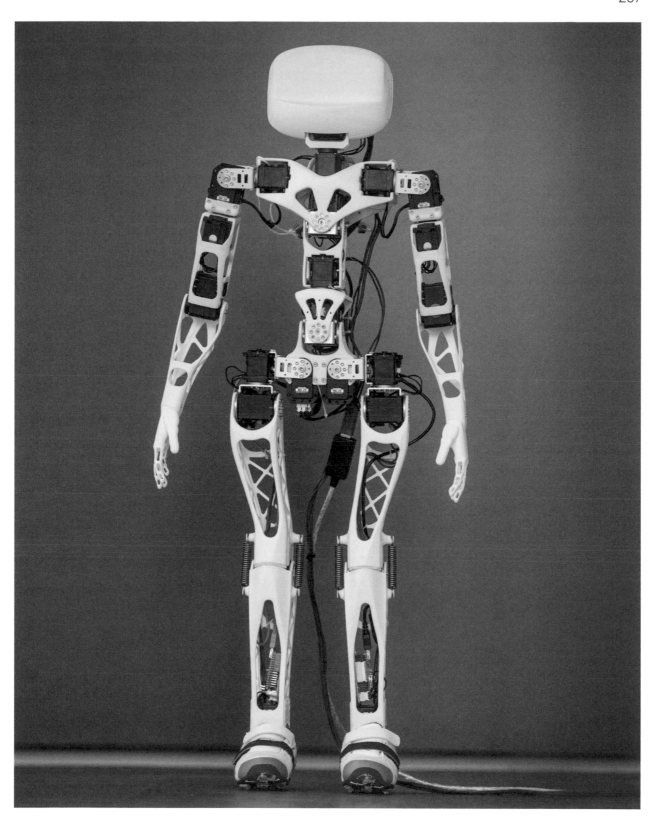

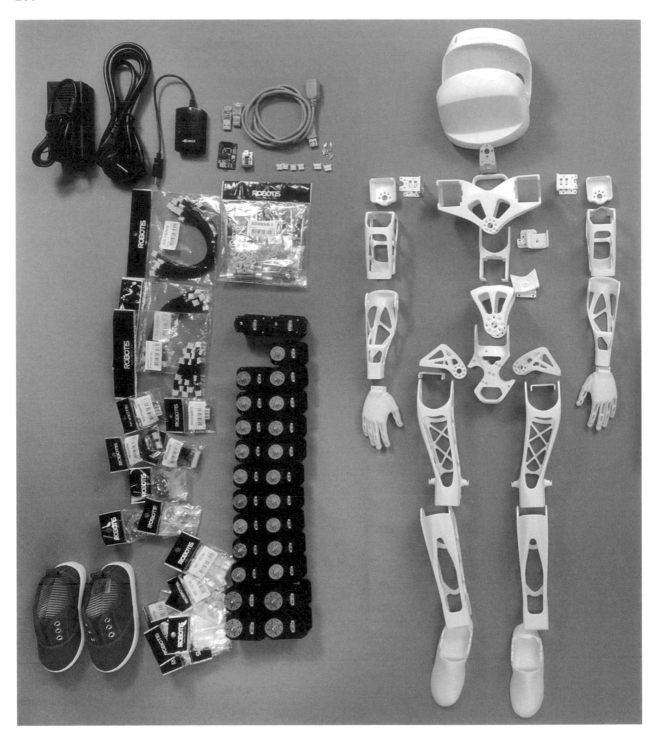

Poppy has relatively few parts for a fully
working robot, making its assembly
relatively easy.

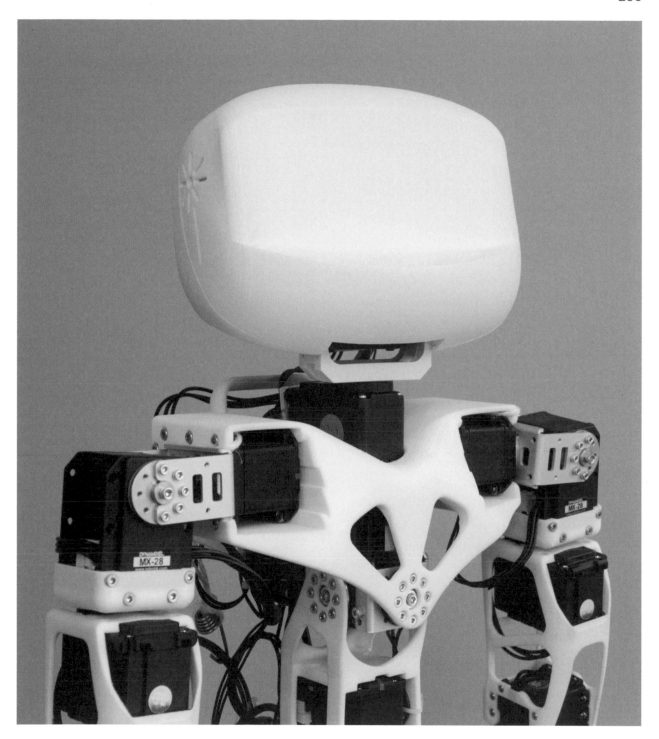

Poppy's head is still very simple, as more attention was paid to the robot's movements. But since this is an open source project, the Fab Lab is free to add more detail and expression to the face.

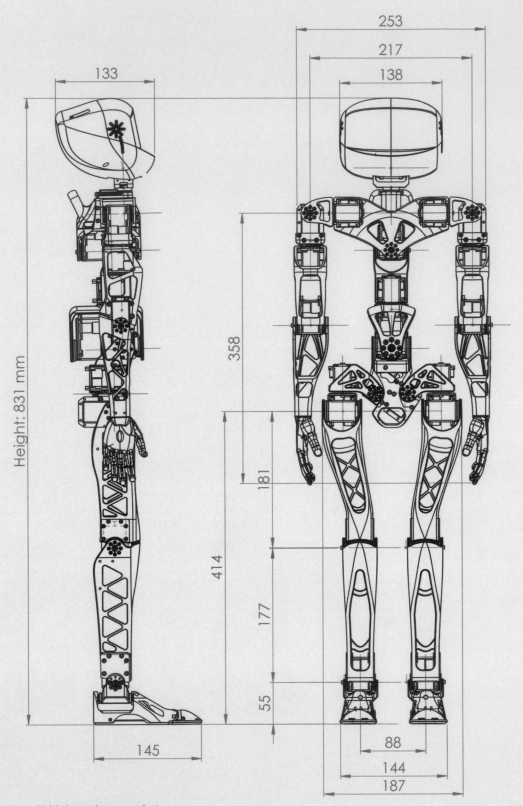

Height: 831 mm

133

253
217
138

358

181

414

177

55

145

88
144
187

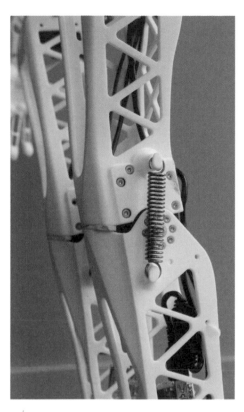

All the joints and mechanisms of the Poppy robot were designed with its movements in mind. It is still a simple robot, but since it is open source its development could be very fast if enough Fab Labs became involved in the project.

AUTHOR.

Primo

PROJECT.

Primo's Cubetto Play Set

#EDUCATION #ENTERTAINMENT

DATE.
2012

DESIGNED/FABRICATED.
DESIGN: SUPSI (LUGANO, SWITZERLAND)
FIRST PROTOTYPE:
FABLAB LUGANO (LUGANO, SWITZERLAND)
PRODUCT DEVELOPMENT:
FABLAB TORINO (TURIN, ITALY)

TECHNOLOGIES USED.
LASER CUTTING
ELECTRONICS PRODUCTION
MICROCONTROLLERS

LINK.
HTTP://PRIMO.IO/

TEAM.
MATTEO LOGLIO (DESIGNER, CO-FOUNDER)
FILIPPO YACOB (CO-FOUNDER)

Cubetto Play Set by Primo is a tangible interface designed to introduce programming logic to 3- to 7-year-old children. It is based on a simple game that involves sending commands to a diminutive robot called Cubetto. The project consists of three parts: an Interface Board, the Cubetto robot and a set of Instruction Blocks. Children create sequences (simplified programs) by placing Instruction Blocks into the holes in the Interface Board. The robot follows the sequence they design, and the children will have to modify the program (the sequence) if the robot doesn't work as expected. The prototype is made out of 4 mm laser cut plywood and two Arduino boards, one for the Interface Board and one for Cubetto. The first prototype was made in 2012 at FabLab Lugano by Matteo Loglio, as part of the MAInD – Master of Advanced Studies in Interaction Design – at SUPSI. The project was developed during the Designing Advanced Artifacts course given by Massimo Banzi from Arduino slongside Habits studio (Innocenzo Rifino and Diego Rossi). With further development at Fablab Torino, Cubetto Play Set is now a product made by Primo.io, a London-based hardware and software company, after a very successful crowdfunding campaign on Kickstarter. The commercialized version of Cubetto Play Set will differ from the prototype, having been given custom design and electronic boards. The company nevertheless provides all the documentation needed for Fab Labs to fabricate the prototype as a Maker version. Both the commercial and the prototype version make use of open source software, hardware and design.

The play set consists of three different components: an Interface Board, the Cubetto robot and a set of Instruction Blocks.

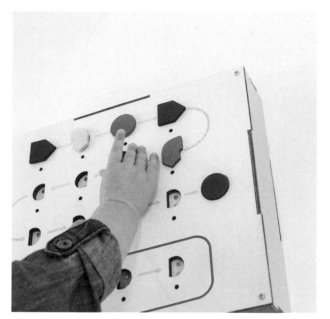

Version 1.0 of the project is a low-cost device that can be manufactured in any Fab Lab. Further versions are going to be more refined, with industrial components and processes, although the team will always provide open source documentation.

AUTHOR.

Enrico Bassi
Maurizio Mion
Andrea Patrucco
Gualtiero Tumolo

DATE.
2014

DESIGNED/FABRICATED.
DESIGNED AT FABLAB TORINO (TURIN, ITALY)
FABRICATED AT FABER SUM (TURIN, ITALY)

TECHNOLOGIES USED.
BIG CNC WOOD ROUTER
LASER CUTTING

LINK.
HTTP://ROOTLESS.TK/

PROJECT.
Rootless

#SPORT #TRANSPORTATION

Fab Labs are important not only as sites where people have access to digital fabrication technologies and experts, but also for their capacity to build networks with other workshops, factories and laboratories in their locality. These connections can be useful for expanding access to expertise, knowledge and digital fabrication technologies that the Fab Lab in question has not yet integrated directly.

The Rootless project provides a representative example of this tendency, having been designed at Fablab Torino and then manufactured at another local workshop, Faber Sum. Rootless is a bike that has all the qualities of a finished product, rather than just a prototype. It was designed with custom and new software and created using a traditional material, wood.

The entire frame is made using a big CNC wood router, and a laser cutter for smaller details. The bike was created from different layers of wood (ash and cherry), with lighter interior layers and more resistant external ones. The various layers of wood are glued together and then milled into two shells that are in turn glued together later, forming one object. In addition, some elements are milled in aluminum, and the team is working on developing 3D printed accessories.

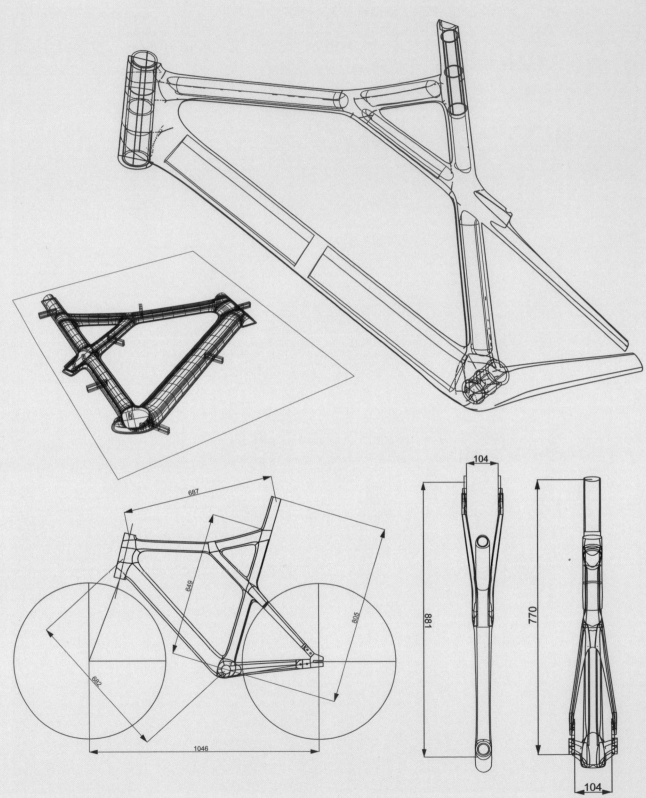

The bike is made using different layers of wood: the inner ones are lighter while the external ones are more resistant.

The whole manufacturing process is quite complex, given the technical requirements of the design. Expertise in the use of a big CNC wood router was a key element in the manufacturing.

The frame is made of many wood layers that are glued and pressed together. This choice is important for the manufacturing process as well as the aesthetics of the object.

This project shows how even professionals can work with Fab Labs and other digital fabrication facilities on very complex projects.

AUTHOR.

Various

PROJECT.

Smart Citizen

#DATA #NATURE

DATE.
2010

DESIGNED/FABRICATED.
DESIGNED AND PROTOTYPE AT
FAB LAB BARCELONA (BARCELONA, SPAIN)
MANUFACTURED IN CANADA, CHINA,
BARCELONA

TECHNOLOGIES USED.
ELECTRONICS PRODUCTION
MICROCONTROLLERS
LASER CUTTING
3D PRINTING: FUSED DEPOSITION MODELING
(FDM)

LINK.
HTTP://SMARTCITIZEN.ME
HTTPS://GITHUB.COM/FABLABBCN/
SMART-CITIZEN-KIT

TEAM.
FAB LAB BARCELONA, INSTITUTE OF ADVANCED
ARCHITECTURE OF CATALONIA (IAAC), MEDIA
INTERACTIVE DESIGN IN COLLABORATION WITH
HANGAR, GOTEO
TOMAS DIEZ, ALEX POSADA, GUILLEM CAM-
PRODON, M.A. HERAS, ALEXANDRE DUBOR,
LEONARDO ARRATA, XAVIER VINAIXA, GABRIEL
BELLO-DIAZ

Fab Labs play an active role in the life of their locality, both through the community that they are the center of, and through the projects they enable and coordinate. A very good example of a project that brings together members of the locality is the Smart Citizen, an environmental monitoring system based on a citizen-led approach to the bottom-up building of Smart Cities. It is premised on the active involvement of 'smart' citizens, rather than a predefined top-down approach. It is critical that the quality of the environment be evaluated and communicated; this process, however, is often only partly enabled by government initiatives, as it would be too expensive to construct a widely distributed network of sensors. Smart Citizen tries to tackle this problem by enabling citizens to build such distributed networks from the bottom-up, starting with the basic unit of the home and expanding to cover whole localities and cities.

Developed collectively at Fab Lab Barcelona, after three years of development it was launched with a crowdfunding campaign on Goteo.org (2012) and then another on Kickstarter (2013). This project is a great example of how products can be developed, prototyped and incubated in a Fab Lab, and then manufactured at a larger scale, using a custom supply chain.

Smart Citizen enables citizens to produce, publish and share their own data. It is based around three components: a hardware platform, an online platform and a mobile app. The hardware platform consists of two interconnected electronic boards: an Arduino-compatible, data-processing board, and a shield filled with sensors capable of analyzing light, sound, temperature, humidity, CO, and NO_2 levels in the local environment. The board is optimized for low power consumption, allowing it to be placed for short timeframes on balconies and windows without being connected to a power source. For longer periods, the board can also be powered by a solar panel. There is also laser cut enclosure (for interior use) or a 3D printed one (with FDM or other technologies) for protecting the board from the weather.

The online platform receives the data from all of Smart Citizen boards in use, then stores and visualizes it. All the data is shared and can be accessed as Open Data with a specific API (Application Programming Interface), software that lets users access the data from the platform automatically and thereby create new applications. Thanks to the API, the system also has an iPhone app (an Android version is being developed) which enables users to manage the board from their smartphones.

This complex and highly detailed system showcases Fab Lab Barcelona's ability to organize, develop, launch and manage demanding and intricate projects. This platform enables citizens to reconnect with their community and environment. In addition, the fact it is open source and open data means that it is modified and owned by them.

The Smart Citizen board integrates a set of sensors and of technical elements that monitor the quality of the surrounding environment.

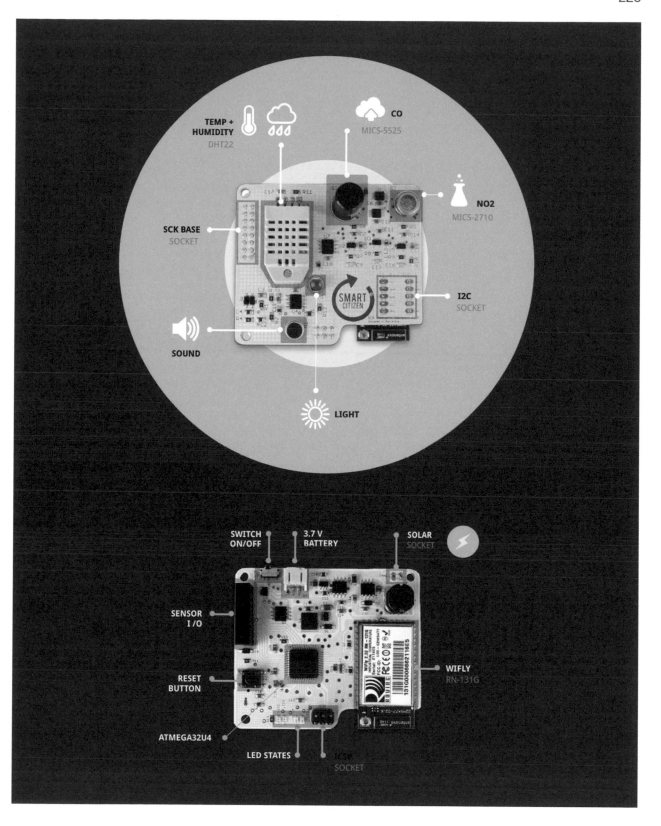

TEMP +
HUMIDITY
DHT22

CO
MICS-5525

NO2
MICS-2710

SCK BASE
SOCKET

I2C
SOCKET

SOUND

LIGHT

SWITCH
ON/OFF

3.7 V
BATTERY

SOLAR
SOCKET

SENSOR
I/O

WIFLY
RN-131G

RESET
BUTTON

ATMEGA32U4

LED STATES

ICSP
SOCKET

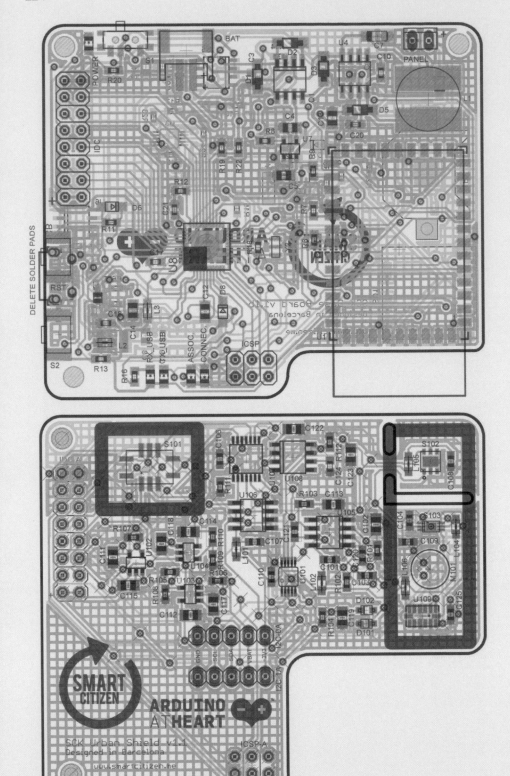

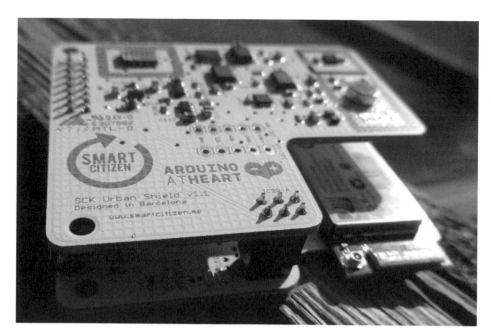

The Smart Citizen board is Arduino-compatible: anybody can modify its software with the Arduino libraries and software, which simplify the development process.

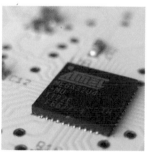
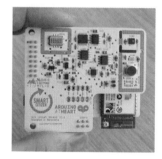

The Smart Citizen board consists of two different boards: a base one for the more technical functions, and an ambient one with all the sensors for measuring the quality of the environment.

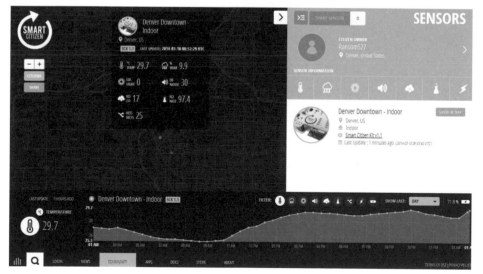

AUTHOR.

Alex Schaub

PROJECT.
Tongue Drum

#MUSIC #ENTERTAINMENT

DATE.
2013

DESIGNED/FABRICATED.
FAB9 (YOKOHAMA, JAPAN)

TECHNOLOGIES USED.
LASER CUTTING
ELECTRIC SAW

LINK.
HTTPS://WEB.ARCHIVE.ORG/
WEB/20150827043426/HTTP://FABLAB.WAAG.
ORG/PROJECT/TONGUE-DRUM

PHOTOS.
GOSUKE SUGIYAMA @ FAB9

Every year since 2005 the worldwide Fab Lab community has gathered at an international conference. Each year a different continent is chosen in order to reach all the local communities. The most recent, FAB10, was held in Barcelona (Spain) in 2014; the next, FAB11, will be organized at MIT in Cambridge (USA) in 2015. In 2013 the FAB9 conference was held in Yokohama (Japan), and, as often happens at these conferences, as well as for meeting and discussing, there was a space provided for making. This was a Super Fab Lab with many different machines, tools and materials; there was also a competition called 'The Open (Re)Source Jazz Orchestra', a contest to build low-cost and open design musical instruments with recycled materials using digital fabrication and other technologies available in Fab Labs. At the end of FAB9, the participants played the developed musical instruments for the whole audience. One of the best instruments created during FAB9 was the Tongue Drum, developed by Alex Schaub from Fablab Amsterdam, a drum made out of a wooden box with 12 tongues, able to emit 12 tones. The design

of the tongues was taken from existing available designs, in order to get the proper sound from the drum. Everything was made out of alder wood: a strong wood is needed, given that the object is a percussion instrument, and thus needs to be resistant and durable. The wooden plate is 18 mm thick, and it was it was not possible to cut it through with the available laser cutting machines. A 18 mm thick plate would need either an adequate laser power or should be able to withstand multiple laser passes (but in this way it could be burnt very easily). The design was therefore laser engraved, and the cutting was done with a manually-operated electric tablesaw. This is a great example of how users in Fab Labs often combine digital fabrication and handwork tools and processes in order to get the best results: even digital fabrication is never completely automatized, as theres always some handwork involved in the manufacturing, assembling or finishing processes. Furthermore, this project is an interesting example of how even temporary Fab Labs can be used to create very functional projects in a very short time.

A picture of the manufacturing process of the Tongue Drum, showing the use of the electric tablesaw on laser engraved paths.

AUTHOR.

Martijn Elserman
Erik de Bruijn
Siert Wijnia

DATE.
2011

DESIGNED/FABRICATED.
PROTOSPACE (UTRECHT, THE NETHERLANDS)

TECHNOLOGIES USED.
LASER CUTTING
ELECTRONICS PRODUCTION
MICROCONTROLLERS

LINK.
HTTPS://WWW.ULTIMAKER.COM
HTTPS://GITHUB.COM/ULTIMAKER/
ULTIMAKERORIGINAL

PROJECT.

Ultimaker Original

#MACHINE

With the recent development of the open source 3D printers of the RepRap project, and the expiry of several 3D printing patents, we have witnessed an explosion of open source and/or low-cost FDM 3D printers. There are so many open source, DIY or commercial versions that it is very difficult to pick one 3D printer as an example, but the Ultimaker is probably the best choice in our case, since it was developed in a Fab Lab and is one of the most commercially successful companies among the many FDM 3D printers.

Martijn Elserman, Erik de Bruijn, and Siert Wijnia organized few workshops at Protospace, the Fab Lab in Utrecht, in order to build some RepRap 3D printers. After these workshops, they decided to design their own 3D printer, inspired by the RepRap, but not necessarily self-reproducible and easier to build in a Fab Lab. The electronics are based on an Arduino board and the frame is laser cut plywood. In March 2011 they launched their project, Ultimaker, with a company of the same name. Initially the printer was only sold as a kit that the users had to assemble themselves. After a while, a pre-assembled Ultimaker was released.

then in 2013 the company released a new version called Ultimaker 2, made of metal sheet and sold pre-assembled, and the first model was renamed (Ultimaker Original). In 2014 the Ultimaker Original was upgraded to include a heated bed, which is the technical component needed to 3D print ABS plastic as well as PLA: this model is now called Ultimaker Original+. The Ultimaker project has always been released under a Creative Commons license, and the manufacturing has always been done by the Ultimaker company and not in Protospace. Here the Fab Lab was useful in the development of the project, which can also be fabricated in any Fab Lab, but that is mostly manufactured and commercialized by an external company in the Netherlands. In September 2014, the company also started manufacturing 3D printers in the USA through a partnership with a local company, a further sign of the success and importance of Ultimaker in the Fab Lab and Maker communities.

A close-up of the print head of the Ultimaker Original. Parts of the print head
are laser cut.

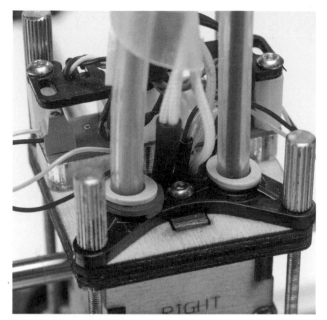

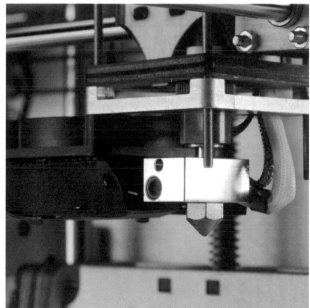

A device with a LCD screen and an
SD card reader was later designed for
the Ultimaker, to let users control the
printer without a dedicated computer.

Even mechanical components are made
of laser cut plywood in the Ultimaker
Original, making it very easy to manufacture in a Fab Lab.

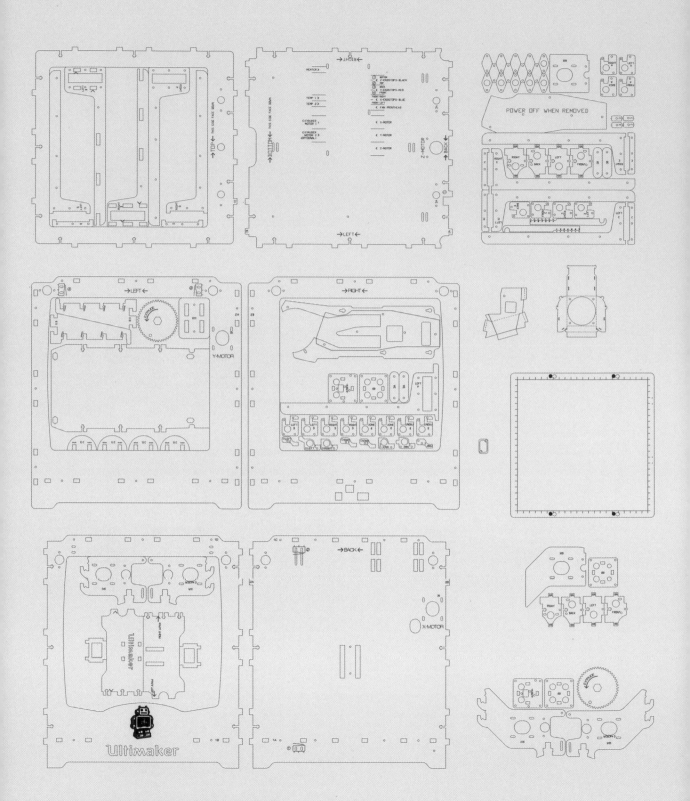

The filament spool is held on the back
of the 3D printer.

AUTHOR.

Denis Fuzii
(Studio dLux)

PROJECT.
Valoví Chair

#FURNITURE

DATE.
2013

DESIGNED/FABRICATED.
DESIGNED IN SÃO PAULO (BRAZIL),
FABRICATED GLOBALLY

TECHNOLOGIES USED.
BIG CNC WOOD ROUTER

LINK.
HTTPS://WWW.OPENDESK.CC/STUDIO-DLUX/
VALOVI-CHAIR
HTTP://WWW.STUDIODLUX.COM.BR/PORTFO-
LIO/VALOVI/

Opendesk is becoming one of the leading online platforms for open design or, more generally, Creative Commons Licensed Design (Creative Commons licenses are not open source if they forbid commercial use and modifications), especially for furniture design or for projects that require big CNC wood routers. Opendesk is based in the UK, but it is available worldwide for uploading, downloading and fabrication of design projects. One of the nicest example is the Valoví Chair, designed in São Paulo (Brazil) by Denis Fuzii, but available and fabricated globally in India, the UK, Australia and Spain, in Fab Lab Leon, for example. The chair can be fabricated with a big CNC wood router from a single 1220 mm x 2440 mm sheet of 15 mm plywood or MDF (different sheets of different colors can be used, and are suggested by the designers). All the pieces can be assembled from a flatpack to a full chair very quickly and without any glue or screws, showcasing that different kind of designs can be implemented with the same technology.

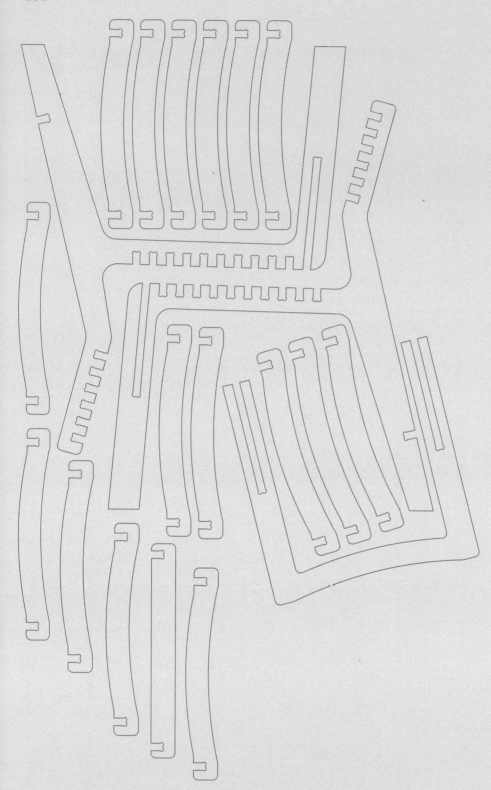

The manufacturing process of this chair is very simple: it requires only a big CNC wood router and one sheet of wood.

The chair can then be easily assembled by hand, and therefore easily transported.

The fabricated and assembled chair in all its glory.

AUTHOR.

Afate Gnikou

PROJECT.
W.Afate

#MACHINE

DATE.
2013

DESIGNED/FABRICATED.
WOELAB (LOMÉ, TOGO)

TECHNOLOGIES USED.
ELECTRONICS PRODUCTION
MICROCONTROLLERS
PRODUCT HACKING

LINK.
HTTP://WWW.WOELABO.COM/WAFATE
HTTP://FR.ULULE.COM/WAFATE/

Digital fabrication, making and Fab Labs can have a real impact everywhere, even places with limited resources, thanks to the power of creativity combined with open access to projects and technologies. A very good example of this comes from Africa: the W.Afate 3D printer. The name comes from the fusion of the 'W' of WoeLab (the Fab Lab of Lomé, capital of Togo), and 'Afate' from the name of the inventor, Afate Gnikou. At WoeLab, Afate worked with a RepRap 3D printer imported from Europe, but soon discovered some limitations, for example, regarding the availability of components in Africa. Afate, still inspired by the RepRap project, then developed his own 3D printer by assembling components that he found in the e-waste dumping grounds close to Agbogbloshie (Ghana): a computer was reused as the frame for the printer, for example, while old scanners were reused as motors and belts. He created a very low-cost machine that recycles the toxic components that are, disgracefully, shipped to Africa for disposal, without proper storage or management.

PLA and ABS plastic filaments and specific components still have to be imported in Togo, and this may not be the complete solution for making digital fabrication accessible in Africa. It is nonetheless a great step forward that showcases how creativity can adapt and make local any technology. Such projects also show us that we should rethink many assumptions about the use of technology for local economic development, even for

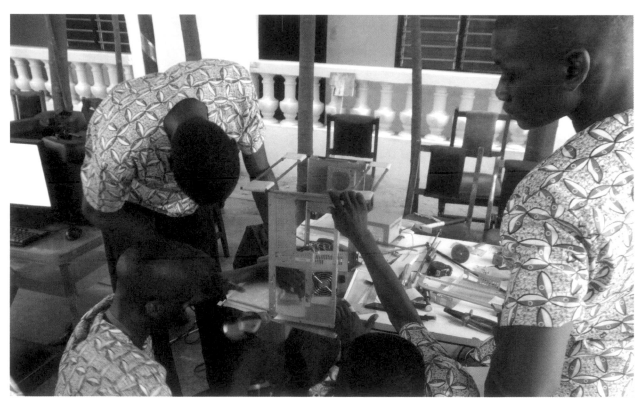

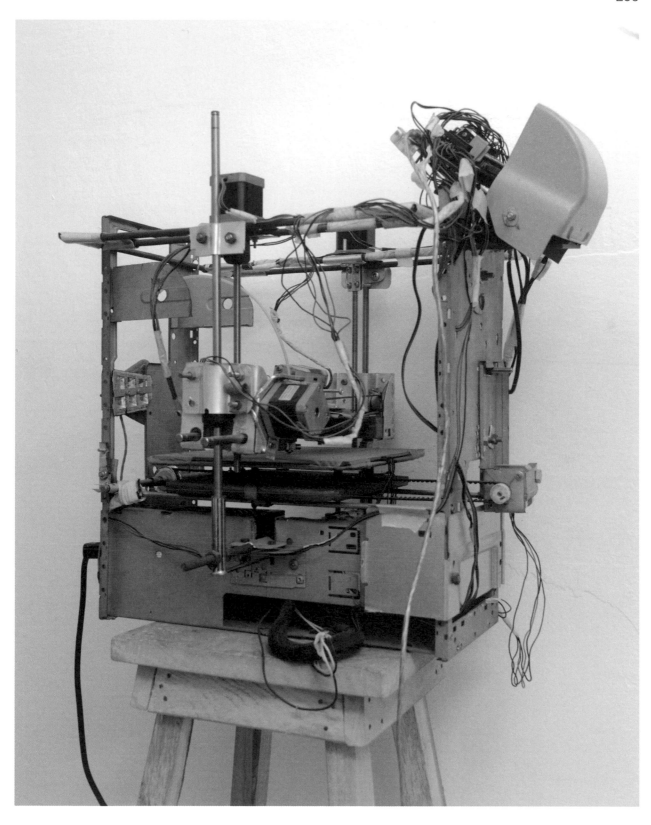

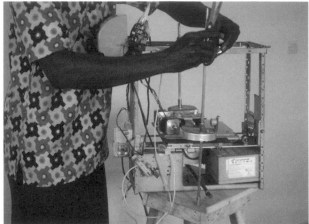

While many projects developed in Fab Labs require components and materials that are produced and shipped from far away, this printer is made out of components found in the local e-waste dumping sites.

Furthermore, this printer shows how it is possible to create digital fabrication machines with just a few tools, very cheap and reused components, and expertise and creativity.

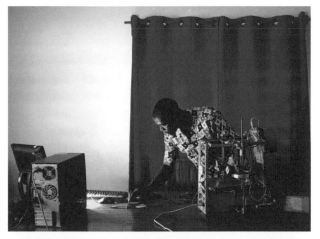

Above images: ©Daniel Hayduk.

the implementation of Fab Labs: projects cannot be implemented everywhere in the same way as in Europe or USA. We need to empower users and give them access to tools, technologies and the knowledge needed to make such projects fit for their locality. It is a longer and more expensive process, but it is critical in such contexts that are so different from Europe and USA. Furthermore, the existence of a Fab Lab in these localities means not only that we can share European or American knowledge with African, South American or Asian Fab Labs, but that also Europe and the USA can learn from these continents. And it is here that having a global network of Fab Labs becomes important, for real peer-to-peer interactions between labs and not just one-way or top-down interactions. Furthermore, such projects (and their limitations) show us that many of the problems that have to be addressed are related to the redesigning of global supply chains of components and materials, in order to make them shorter and more sustainable, both for manufacturing and recycling.

AUTHOR.
Cesare Griffa Architetto
(various)

PROJECT.
WaterLilly 1.1 / LillyBot 2.0

#MACHINE

DATE.
2013

DESIGNED/FABRICATED.
FABLAB TORINO (TURIN, ITALY)

TECHNOLOGIES USED.
ELECTRONICS PRODUCTION
BIG CNC WOOD ROUTER
LASER CUTTING
ELECTRONICS PRODUCTION
MICROCONTROLLERS

LINK.
HTTP://CESAREGRIFFA.COM/WATERLILLY/
HTTP://CESAREGRIFFA.COM/LILLYBOT-2-0/

TEAM.
WATERLILLY 1.1: CESARE GRIFFA, MASSIMILIANO
MANNO, DENISE GIORDANA, FEDERICO BORELLO
LILLYBOT 2.0: CESARE GRIFFA, MATTEO AMELA

Many projects developed in Fab Labs are machines. Many are for digital fabrication, but even more experimental machines are also developed. One such is the Water Lilly family, a series of interactive devices that act as photobioreactors to grow micro-algae. These devices were designed as architectural components, with a view to introducing the cultivation of micro-algae into houses and buildings and making it a much more common practice than it is now. Experimentation becomes an every-day activity. Such organisms are studied today as one of the most promising sources of bio-fuels, polymers or food that can

be produced with just light, mineral salts and CO_2. These devices integrate the production of carbon based polymers with the absorption of CO_2 from the environment, the production of O_2 and air and water filtering. With the introduction of Arduino boards and sensors, these devices interact with people's presence and movement and can gather data about the environment and the cultivation of micro-algae. The structures of these devices is mainly built out of laser cut and engraved plywood or plexiglass (PMMA), while bigger components are CNC milled.

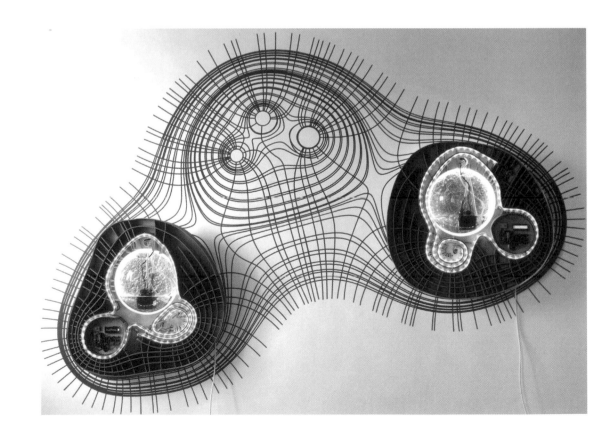

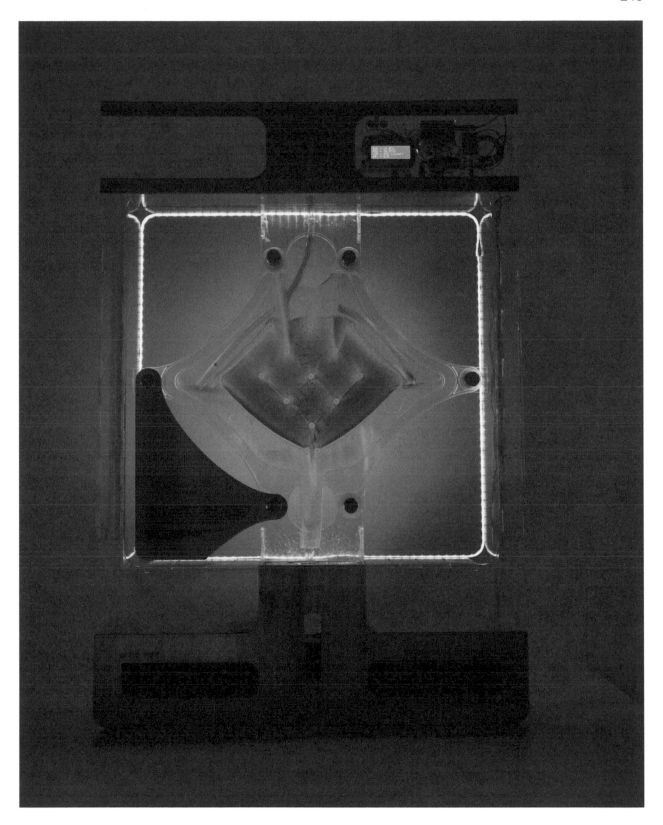

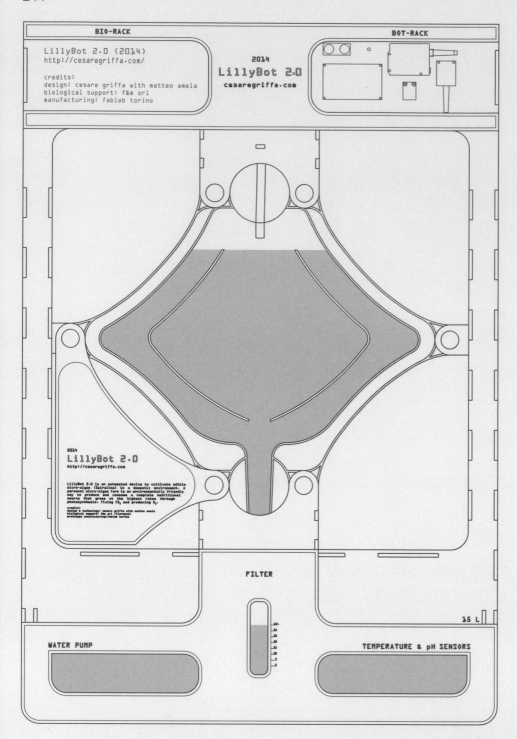

BIO-RACK

BOT-RACK

LillyBot 2.0 (2014)
http://cesaregriffa.com/

credits:
design: cesare griffa with matteo amela
biological support: f&m srl
manufacturing: fablab torino

2014
LillyBot 2.0
cesaregriffa.com

2014
LillyBot 2.0
http://cesaregriffa.com

LillyBot 2.0 is an automated device to cultivate edible
micro-algae (Spirulina) in a domestic environment. A
personal micro-algae farm is an environmentally friendly
way to produce and consume a complete nutritional
source that grows at the highest rates through
photosynthesis fixing CO_2 and producing O_2.

credits:
design & technology: cesare griffa with matteo amela
biological support: f&m srl (florence)
prototype manufacturing: fablab torino

FILTER

15 L

WATER PUMP

TEMPERATURE & pH SENSORS

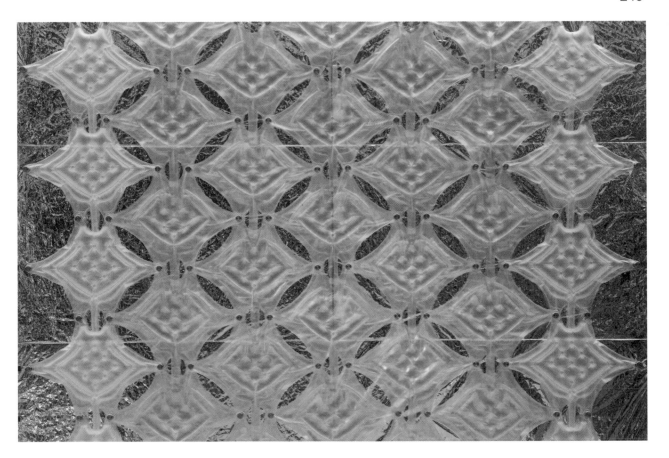

The containers of the micro-algae
can have different designs.

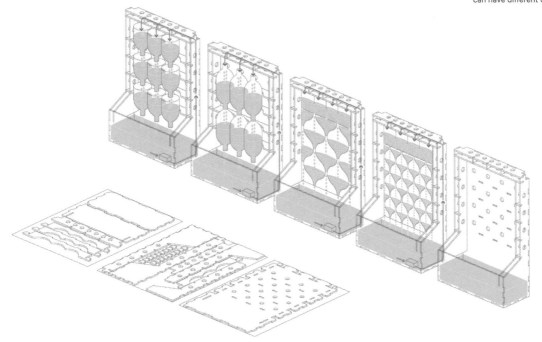

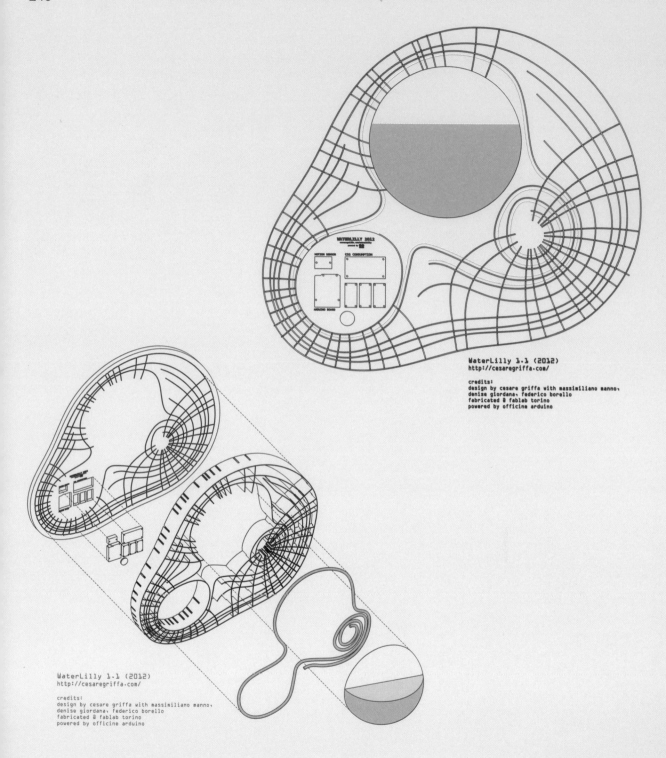

WaterLilly 1.1 (2012)
http://cesaregriffa.com/

credits:
design by cesare griffa with massimiliano manno,
denise giordana, federico borello
fabricated @ fablab torino
powered by officine arduino

WaterLilly 1.1 (2012)
http://cesaregriffa.com/

credits:
design by cesare griffa with massimiliano manno,
denise giordana, federico borello
fabricated @ fablab torino
powered by officine arduino

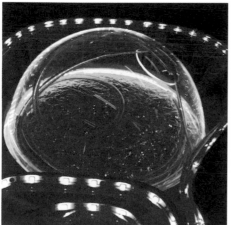

This project shows how such a technical device could be used in houses and as an architectural element.

Heloisa Neves

Pioneers

Global concepts in local communities

Being a pioneer means being the first person to implement an idea or a project in a given field. Being a pioneer within the Fab Lab universe means going even further. 'Fabbers' not only implement the first Fab Labs in their regions: they help the local community to become familiar with technology and new ways of working, based on innovative and often unfamiliar concepts such as peer learning, open process, democratization of technology and knowledge sharing. In other words, they build a structure from nothing, managing local partnerships for funding, learning techniques related to digital fabrication processes, and, finally, convincing others to share their vision of the benefits that a space like this could bring to the local community.

This chapter serves as a brief introduction to some of the pioneers of the Fab Lab network. It will attempt to provide an insight into the importance of these people, not only for Fab Labs, but within a new movement that has been gaining ground in the world, which some refer to as the Maker ecosystem. It is based on a 'hands on' philosophy, on the pleasure and satisfaction of 'doing it yourself' and 'doing it with others', and in particular on the pursuit of making technology a more human and democratic process.

As is well-known, the Fab Lab network began in the United States under the guidance of Neil Gershenfeld, the first great pioneer and the man responsible for making the network possible. Neil likes to say that the Fab Lab network began almost by chance, that it sprung up inspired by an intuitive understanding, and by a desire to bring into being a more open and collaborative mode of working; today, it constitutes a new way of thinking about products, services and processes, in an environment of constant discovery and innovation.

Neil made reality something that, on paper, seemed improbable: he managed to sell the idea that it would be interesting to create a space with 'strange' machines that could 'do almost anything'. A rather vague, ambitious concept, perhaps, but less so than the one that was his real dream: to create a more human and collaborative world through technology.

Neil thus had the audacity to implement the first Fab Labs by way of an unusual process: he did not present a concrete business model based on building an international network, nor did he promise his partners that in ten years they would reach the target of opening more than 300 laboratories around the world (a total which has not only been reached, but surpassed!). Neil presented his partners and future pioneers with real stories, with projects involving those children and adults who came to the first Fab Labs and, simply, fell in love with being there, with learning and sharing knowledge in an unusual and fun way.

Many other pioneers have followed Neil, and each one has sewn together another piece of this never-to-be-finished patchwork quilt. It is always in process. These pioneers devoted their efforts to helping the idea grow and mature. This was the initial role of the pioneer, to give support to a battalion of other Fabbers, who in turn contributed to enriching the spaces with creativity and interesting projects.

Personally, I remember that when I first started becoming familiar with the Fab Lab universe, I wondered who the minds behind them were. It seemed to me that the authors had to have great powers of understanding, of innovation, of recognizing potential and bringing together communities. I remember that I was also very impressed with the horizontal and collaborative management of the first laboratories I visited. Fab Lab Barcelona was the first. I had an informal meeting with Tomas Diez (director of the Fab Lab Barcelona); I went to the lab and came across a range of excellent projects and many machines with which I was unfamiliar. I expected to meet a rather staid person, who could tell me a little about digital manufacturing in a traditional sense. And then Tomas arrived, a young man full of ideas of how to connect people to their cities, and empower them to construct something together. Tomas' perspective and vision had a memorable impact on me.

It was Benito Juárez (current director of the Fab Lab Peru and Fab Lat) who put me in touch with Tomas. Benito is an old friend, and perhaps the most important figure in my personal Fab Lab trajectory, since he was the person who introduced me to the network. Benito's ideas (how to make the world more beautiful and more human) seemed fantastic to me, almost fanciful. But standing there in the Fab Lab Barcelona, I began to see what he was talking about. I began to share the vision of Tomas and Benito, and see that this could be the place to construct something truly significant.

These were the first two pioneers that I met; there have been so many others. Each has reaffirmed the picture that I began to develop the first time that I went to Fab Lab Barcelona, each one in their own different and brilliant style.

To continue with the Spanish pioneers: I had the honor of working with Jose Perez de Lama (Osfa) on one of the most collaborative laboratories of the network, Fablab Sevilla. The atmosphere created by Osfa is very homely, almost like a family; a place to learn with friends. I would also like to mention the work of Nuria Robles from Fab Lab León, which has provided the foundation for a strong and important network of cooperation and learning. Moving on to Holland: Alex Schaub is one of the most sophisticated digital artisans I know, not only because he works on the border between traditional technique and its connection with the digital, but also because he fills the Fablab Amsterdam with music of the highest quality. Joris van Tubergen, another pioneer, expanded the ideas of being a designer and of creating something new, making them more open and collaborative. And, still in the Netherlands, I could not fail to mention Bart Bakker, the creator of the first Mini Fab Lab in the garage of his home, one of the people who best understand collaboration and sharing dreams.

I must also mention other pioneers who have influenced me, each in their own way: Nicolas Lassabe, Laurent Ricard and Emmanuelle Roux from France, who put into practice the dream of creating a social place connected to the reality of the community. Also the Italian Massimo Menichinelli, editor of this book and the man responsible for turning projects and processes carried out in the network into words and ideas. And also Kamau Gachigi, from Nairobi Fab Lab, the person in charge of spreading African innovation through Fab Labs and inspiring people, sometimes against the odds, to fabricate a better life.

The more I write, the more names start to pop into my mind: the first Fab Nomad, Jens Dyvik, Victor Freund from Peru and, definitely, the 'mother' of the Fab Lab network, Sherry Lassiter, from Fab Foundation, who helps all the other pioneers and collaborators to expand the potential of their laboratories and make them truly a place of learning and friendship. And finally I would like to mention the name of Paulo Fonseca, who believed in the idea of a Brazilian laboratory and echoed my desire to develop a community in Brazil dedicated to realizing the potential of digital fabrication as a means of social development.

I have mentioned all these many names in order to give the reader a few leads for stories to be followed, and as a reflection of my own personal engagement with the Fab Lab network. But despite this profusion of names, I have mentioned only a tiny portion of all the pioneers in the network.

But this text is not solely about my own experiences and personal vision of the network. In preparing it, I asked four pioneers from four different corners of the world to share some of their experiences. I believe that their voices will give a sense of the experience of all of the network's pioneers.

The first is **Frosti Gislason**, from Fab Lab Vestmannaeyjar in Iceland. Talking to Frosti is an opportunity to learn a little more about how this network started, and as a laboratory succeeded in uniting local traditions with global knowledge.

I also spoke with **Hiro Tanaka**, responsible for the first Fab Lab in Japan and currently responsible for the Asian network. Hiro shared many lessons about how to help people find their own answers through 'invisible' facilitation.

Another pioneer who shared her vision with me was **Wendy Neale**, from New Zealand; she is one of the pioneers who coordinates one of the most distant, but no less connected, laboratories of the network.

And finally, it is a privilege to share the beautiful words of **Benito Juárez**, a pioneer in Latin America and the man most responsible for the expansion of the network in this part of the world.

I will let them speak for themselves.

Frosti
Gislason

FAB LAB VESTMANNAEYJAR

Frosti Gislason is an industrial engineer. He started the first
Fab Lab in Iceland in Vestmannaeyjar in 2008, and has been
the Fab Lab manager there from the start. Frosti has been
leading the Fab Lab Iceland network from within the Innovation Center Iceland; there are now five Fab Labs in Iceland.

Hiro
Tanaka

ASIAN FAB LAB NETWORK
FAB LAB KAMAKURA

Hiro Tanaka is Doctor of Engineering at the Faculty of Environmental Information, Keio University Associate Professor,
Fab Lab Japan Founder, Fab Association Asia representative
and Fab Lab Kamakura Co-Founder.

Wendy
Neale

FAB LAB WELLINGTON

Wendy Neale Is a furniture designer and manager of Fab Lab
Wellington. With a digital and craft-based practice, Wendy
designs and creates meaningful objects from waste and
obsolete furniture, and also develops furniture with modified
traditional joints using digital techniques. She's most excited
when combining the two.

Benito
Juárez

FAB LAB PERU
FAB LAT

Benito Juárez is an architect at the National University of
Engineering. He graduated from the Fab Academy program
in 2009 at the IAAC in Barcelona, Spain, with the mission of
founding the first Fab Lab in Latin America. He is now director of the Fab Lat Latin America Network. He founded the
FABLAB-MET with Victor Freundt, advised on the foundation
of the FabLab-TECSUP, and has helped in the opening of Fab
Labs in Latin America, including FabLab-Cali, FabLab-Mexico
and FabLab-Salvador.

What was it that first drew you to the world of Fab Labs?

FG 'When one of my staff members saw Neil Gershenfeld's TED lecture "Unleash your creativity in a Fab Lab"[1] , the world changed. We decided that a Fab Lab was something we needed in our little town in Vestmannaeyjar, a group of volcanic islands in the south of Iceland. I thought it would be great to have a fab lab in Vestmannaeyjar because of fab lab ideology, and the possibility of knowledge transfer across the globe. I found it fascinating to have access to all these people around the world, and to be able to share knowledge and ideas and make almost anything using digital fabrication methods.'

1• Gershenfeld, N. (2006). Unleash Your Creativity in a Fab Lab. TED Talk. Available at: http://www.ted.com/talks/neil_gershenfeld_on_fab_labs

2• Editor's note: CBA, FAB5. Available at: http://cba.mit.edu/events/09.08.FAB5/index.html

3• Editor's note: Gershenfeld, N., 2005. FAB: The Coming Revolution on Your Desktop-From Personal Computers to Personal Fabrication. Basic Books.

HT 'After I was fortunate enough to be appointed an Associate Professor at the X-DESIGN program in Keio University, Japan (2005), I had been looking for a new research field and new direction in my career. While my background was Architectural CAD and 3D Modeling (I did my PhD in research on image-based 3D scanning in 2003), I was teaching Web Design, Electronics, Media Art, Visualization – (almost) anything related to making. It was fun, but I was struggling to find an identity for myself. I was looking for something which would offer me a way of integrating my scattered interests, something which would make my path clear, something which would give a name to my profession. One day in summer 2008, I stumbled across the website of FAB5 (India)[2]. I had already read Neil's book FAB.[3] It was just one week before FAB5. I felt something special and, without quite knowing why, I attended FAB5. Soon after arriving I realized, "Oh, this is my place! This is what I have been looking for! They are people whom I have been longing to meet!" So, my FAB story started in FAB5 Pune, when I met Neil, Sherry and all my dear friends!'

WN 'Chris Jackson, the instigator of Fab Lab Wellington, was questioning what a post-industrial design era would bring to the world. He was particularly interested in the democratization of design and open source communities. As a result of this interest, he discovered fab labs and was very inspired by the values underpinning them. He attended the FAB7 conference in Lima, and then enthusiastically pitched for FAB8 to be held in Wellington, and we were lucky enough to be successful. A very intense year followed and Fab Lab Wellington was established in time for FAB8, thanks to Chris's inspiration and the work of a fantastic team.'

BJ 'I got involved in this project as a consequence of the fact that my life has been defined by the risks I've taken. I grew up in the Peruvian jungle, where nature inspired in me a special sensitivity and creativity, but where I was also witness to the impact of terrorism on one of the poorest areas of my country. These situations, nature and creativity versus terrorism and exclusion, would mark, years later, my vocation to connect "innovation and inclusion". My family moved to Lima and, at 16 years old, I fell in love with a city that enabled me to connect with a new landscape, with their "forests" of mats and bricks, their "rivers" of vans and vendors, and its gray sky. The city opposed my sense of the jungle's freedom with the extreme need of many families that, like mine, had to migrate to the capital. The city of Lima became my first laboratory, where I experimented with technological projects for social innovation (ovo-sys.blogspot.com). Thanks to these projects, in 2009 I was selected (with my partner Victor Freundt) to be the founder of the Fab Lab Lima (the first in South America), a project of the Center for Bits and Atoms at the Massachusetts Institute of Technology, sponsored by the Spanish Agency of International Cooperation for Development (AECID) and promoted by the Institute for Advanced Architecture of Catalonia (IAAC, Barcelona). I took the risk and emigrated again! After one year in Fab Lab Barcelona, I returned to Peru (2010), motivated to provide access to these technologies to more people, especially those from the most excluded social sectors.'

QUESTION.

In your personal view, what remains to be achieved?

ANSWERS.

FG 'Locally, it will take a little longer to have tangible jobs that we can say come directly from the fab labs. We want to build things that can have a great impact on the lives of people around the globe.'

HT 'Material Recycling: It has been four years since I started the first fab lab in Japan. The number of fab labs in Japan is steadily increasing; people enjoy learning, making and sharing. I am very happy to see that; my dream came true. But I always found a huge amount of "creative mess" after workshops. People left lots of wasted acryl boards, lots of plastics, lots of sheets, garbage, and so on. When someone makes something, he or she also produces a lot of waste. I'd like to solve this problem. Digital Material (or FabLab3.0) is a nice approach. At the same time, we Japanese people have a sense of "MOTTAINAI", which has inspired various recycling and upcycling techniques. I think the Fab Lab is the best platform for building a "techno-ecological" world. I'd like to continue with my research on Material Circulation.'

WN 'At Fab Lab Wellington we are still developing our expertise and our connections with the rest of the network. Establishing an open source space in a university environment has had its challenges, and some still remain, but the development of Fab Lab Wellington has been very beneficial to the University. The network has been great for us because we are physically isolated from much of the world, but we have been able to be more connected through the network. There are very exciting projects that we're involved in and many more that we'd like to be involved in – there never seems to be enough time to do everything. It would also be great to have more Fab Labs in New Zealand and in the South Pacific, especially in remote communities.'

BJ 'Inclusion requires adaptation. In the last four years, the growth of Fab Labs has been exponential, reaching more than 250 laboratories worldwide [in 2013, when this interview was made]. Despite looking to be a globally inclusive project, 75% of Fab laboratories are located in developed countries (40% in Europe and 35% the US), and only 5% in Latin America. What are the causes of this difference? Perhaps the concept of "technological democratization", which is central to Fab Labs in developed countries, is incommensurate with Latin American reality. There are several key factors. One is the notion of "innovative thinking": most universities and institutes in Latin America guide the students towards an education in consumer technology, but not in development. This reality is reflected in professionals, companies and organizations that focus their activities on production and trade but rarely on innovation. Less than 2% of companies incorporate innovation and development activities. I could also cite "cooperative thinking": to the low level of innovation we add an entrenched culture of competition, of win/lose. Most companies talk a lot about cooperation, but they don't practice what they preach. And, to finish, the economic and administrative factor: acquiring skills, tools and digital fabrication equipment in Latin America is between three to eight times more expensive than in Europe or in the US (import expenses, transportation, customs, living costs etc.) In addition, bureaucracy can take between three to six times longer than in the U.S. or Europe. How, then, can we generate technological democratization in Latin America? How can we encourage the development of technologies that allow us to provide greater value to our cultural and natural heritage? How can we take the leap from being a semi-industrial society, to developing processes and products in line with the new industrial (r)evolution?'

QUESTION.

Regarding the history of your lab, can you highlight a few points that make your laboratory a special place?

ANSWERS.

FG 'The Fab Lab in Vestmannaeyjar is special in the sense that it's in a small community and has affected the lives of many people of the town. It has opened the eyes of the inhabitants to new technology and new ideas. The Fab Lab has enabled many people from different backgrounds to work together, and one of our favorite projects involved half of the population of the island working together to make the town a little nicer. The Fab Lab in Vestmannaeyjar has focused on educational projects and has a strong connection with all school levels on the island.'

HT 'When I started my Lab in Japan, nobody understood me. I asked lots of people to assist me setting up the first fab lab in Japan, but I received no official financial support. But I couldn't wait. I decided to buy equipment (laser, 3D printer, paper cutter, milling) with my own money (I had to give up on my idea of buying a car). The next problem was related to location. I thought it was necessary to have a kind of "symbolic" feature, in order to characterize the first Fab Lab. So Youka and I finally found a 150-year-old Japanese wooden traditional building. We decided at first sight to rent the building for the Fab Lab. The atmosphere of traditional Japanese wooden buildings is very different from modern architecture. People are always saying they feel a welcoming and cozy air when they come into our building. Fab Lab Kamakura is open to every generation. Whenever I go there, I find young people and old, men and women, Japanese people and foreigners. I had never seen this kind of place in Japan.'

WN 'Our Lab is special because we are based in an Art and Design School, and have a wide range of interests and expertise to share with our local community and the wider Fab Lab network. We are also special because we are based in Wellington, a fantastic city, full of creativity and collaboration. One of our best projects to date is our Resilience Project, which is looking at ways to become more sustainable – through collaborating with WikiHouse New Zealand, the Open Source Beehives project, a Weta Hotel project, and setting up a waste minimization initiative on campus that is bringing together people who work, study and live near the university.'

BJ 'Today, as president of Fab Lab Peru, Fab Academy International Tutor and director of Supernode[4] Latin America, I work for technological democratization through the creation of new laboratories and cutting edge products based on digital manufacturing, reassessing Latin American biodiversity and multiculturalism. And at our labs we can learn a lot about multiculturalism and eco-Diversity. This is certainly the most exciting action that we have constructed here.'

4. Editor's note: in the Fab Lab community, Supernodes are the most relevant labs in the network, thanks to their history, size and activity.

To end on a more personal note: what makes you happy working with the Fab Lab network?

FG 'It makes me really happy to see the expressions of users when they manage to make what they wanted. I like to see how we can change the minds of young people and help them to improve their self-confidence by making things, knowing how to design things, and programing.'

HT 'Friends. (Almost all) the people connected to the Fab Lab are self-motivated, self-initiated, open-minded and helpful to each other. I am happy to work with nice people.'

WN 'While Fab Lab Wellington was being set up, I was living and working in Portugal. I was telephoned by the University and invited to return to New Zealand to manage the lab. I accepted the invitation and arrived back in New Zealand ten days before FAB8. I have been enjoying my role at Fab Lab Wellington ever since. I love working with people who are inspired and inspiring, who love what they do. For me, the Fab Lab network isn't just about shared digital fabrication: it's about shared enthusiasm and sharing skills and ideas to change the world. I love that we can call on each other for support and guidance and that everybody is keen to learn more about each other's experience and expertise. I love the way that the network is developing to find better ways to support each other, and help new labs to set up. It's exciting that I know people from many countries, can talk to them on the MCU[5], and can work with people in other parts of the world, even though we've never met.'

BJ 'Making ideas and passion real, as in the project Floating Fab Lab on the Amazon River, a Fab Lab Peru project in collaboration with Fab Foundation, Fab Lat, Project A+ and UA Bureau. We face a revolution that is transforming the daily lives of people (digital fabrication) and a territory (the Amazon Rain Forest) that has great potential to respond to the world's challenges. It is the ideal place to incubate the manufacturing processes of the future; to explore alternatives, and work towards a responsible and responsive industry; to integrate local and global processes; to provide access to the benefits of digital manufacturing to the native population, helping to solve problems related to health, energy and education; and to bring together people, institutions and countries around the world in the conservation of the Amazon.'

5• Editor's note: MCU is the videoconferencing system that connects all the Fab Labs worldwide. It can be accessed at http://mcu.cba.mit.edu/

I believe that through the eyes and words of each of these pioneers, the quilt is being formed. There is still a lot to build, but we have already seen a huge amount of results. For me, and certainly for the pioneers, there is always a desire to do better; this chapter is thus more than a portrait of these people: it is an invitation to each of you to help us continue weaving our network and strengthening it with other views and experiences.

Appendix

Fab Labs geography - Authors - Further Reading - Acknowledgments

The Fab Lab network is constantly evolving, with new machines, processes, formats, business models and events being integrated. In this last chapter you can find more data about the global Fab Lab network and a list of publications that will give you further insight into this world. Here too you will find the biographies of the authors of this book as well as references for all the facts cited.

Fab Labs geography

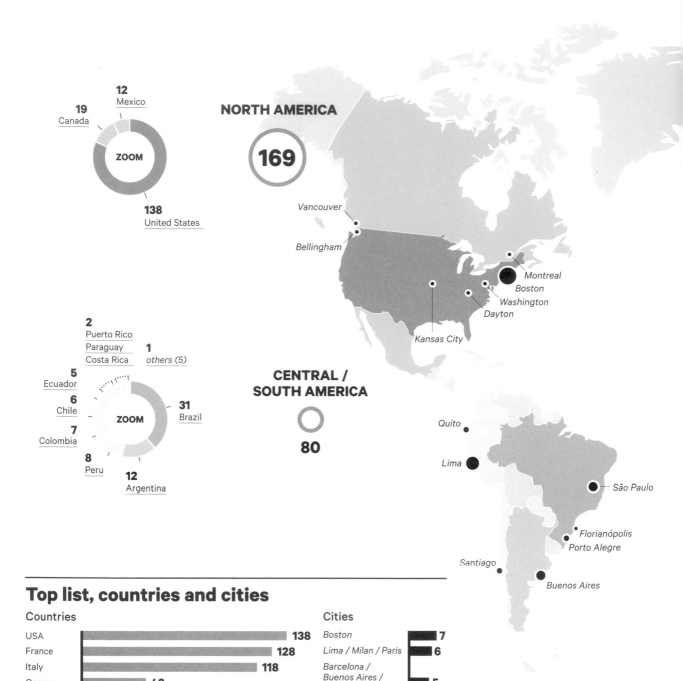

19 Canada

12 Mexico

ZOOM

138 United States

NORTH AMERICA

169

Vancouver
Bellingham
Montreal
Boston
Washington
Dayton
Kansas City

2 Puerto Rico Paraguay Costa Rica

1 others (5)

5 Ecuador

6 Chile

7 Colombia

8 Peru

12 Argentina

ZOOM

31 Brazil

CENTRAL / SOUTH AMERICA

80

Quito
Lima
São Paulo
Florianópolis
Porto Alegre
Santiago
Buenos Aires

Top list, countries and cities

Countries

USA	138
France	128
Italy	118
Germany	43
Spain / UK	37

Cities

Boston	7
Lima / Milan / Paris	6
Barcelona / Buenos Aires / Dubai / São Paulo / Tblisi	5

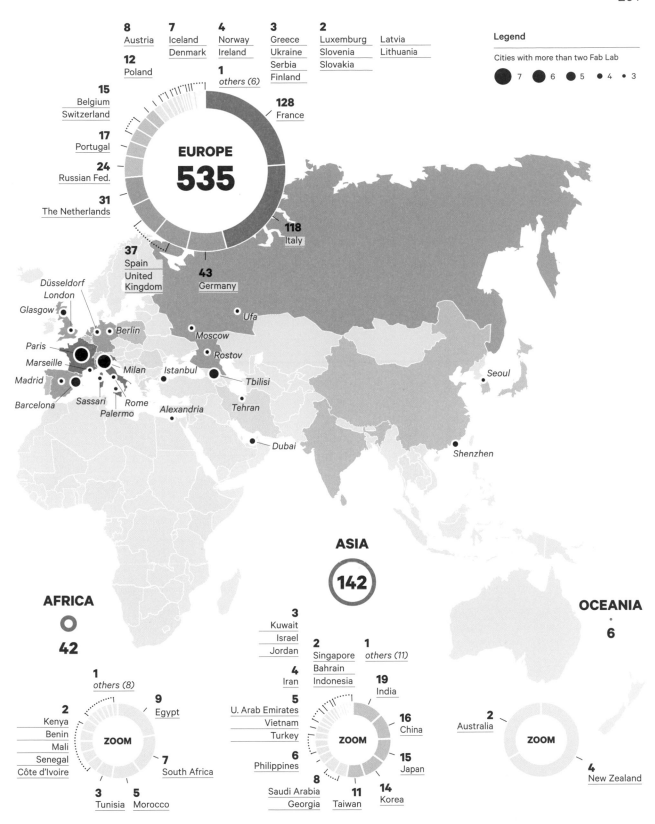

Legend

Cities with more than two Fab Lab

● 7 ● 6 ● 5 ● 4 · 3

8 Austria
7 Iceland / Denmark
4 Norway / Ireland
3 Greece / Ukraine / Serbia / Finland
2 Luxemburg / Slovenia / Slovakia
Latvia / Lithuania
12 Poland
1 others (6)
15 Belgium / Switzerland
17 Portugal
24 Russian Fed.
31 The Netherlands
37 Spain / United Kingdom
43 Germany
128 France
118 Italy

EUROPE 535

Düsseldorf
London
Glasgow
Paris
Marseille
Madrid
Barcelona
Sassari
Palermo
Rome
Milan
Berlin
Istanbul
Alexandria
Tbilisi
Tehran
Dubai
Moscow
Rostov
Ufa
Seoul
Shenzhen

ASIA 142

3 Kuwait / Israel / Jordan
2 Singapore / Bahrain / Indonesia
1 others (11)
4 Iran
5 U. Arab Emirates / Vietnam / Turkey
6 Philippines
8 Saudi Arabia / Georgia
11 Taiwan
14 Korea
15 Japan
16 China
19 India

ZOOM

AFRICA
42

1 others (8)
2 Kenya / Benin / Mali / Senegal / Côte d'Ivoire
9 Egypt
7 South Africa
3 Tunisia
5 Morocco

ZOOM

OCEANIA

6

2 Australia
4 New Zealand

ZOOM

Authors

MASSIMO MENICHINELLI

is a designer who since 2005 has researched and developed open, collaborative, and co-design projects, and the systems that enable them. Holder of an MSc in industrial design from Politecnico di Milano (Milan, Italy), he completed the Fab Academy in 2012. Massimo uses design tools and processes in order to help companies, organizations, cities and local communities to develop open and collaborative processes, businesses, services, places and projects, such as open design, Fab Lab and user-driven open and social innovation initiatives. Massimo has given lectures, talks and workshops in various countries including Italy, Spain, Finland, France, Germany, United Kingdom, Mexico, Colombia, South Korea and Singapore. He worked on the development of the Aalto Fablab and co-organized the first Open Knowledge Festival in Helsinki (Finland). He lectures on digital fabrication and open design at Aalto University (Helsinki, Finland) and open design at SUPSI (Lugano, Switzerland). He has recently developed the MUSE FabLab (Trento, Italy) and facilitated the development of the Opendot Fab Lab (Milan, Italy). He worked as a director at Make In Italy – Italian Fab Lab & Makers Foundation CDB, and he currently works as project manager at IAAC | Fab Lab Barcelona.

CAMILLE BOSQUÉ

lives in Paris. After graduating in product design from École Boulle and École nationale supérieure de Cachan, in 2012 she began working on Fab Labs, hackerspaces and the Maker movement for her PhD. Her research is situated at the intersection of aesthetics, design and anthropology. For her research, she has spent two years visiting makerspaces around the world, to study the implications and applications of personal digital fabrication. As a holder of an 'agrégation' (teaching degree) in applied arts, Camille Bosqué also teaches history of design and design at Université Rennes 2 and Ensci – Les Ateliers, exploring the hidden links between digital decentralized fabrication and historical movements in design history. She has also been leading speculative design fiction workshops with design students, looking at possible futures and prospective scenarios for additive manufacturing. She has published widely on these topics, and recently published Fab Labs, etc. (2014), the first book in French on the subject, which brings together many interviews as well as detailed analysis and descriptions of Makers' practices and makerspaces.

PETER TROXLER

studies the impact of readily available direct digital manufacturing technologies, the design and manufacturing practices of 'fabbers' and 'Makers' in the creative and manufacturing industries, and the emergence of networked co-operation paradigms and business models based on open source principles (such as open design and open source hardware). In addition, Peter studies the emergence of third spaces and new manufacturing initiatives in Urban Open Innovation Environments, how they relocate production and research functions to the centers of neighborhoods in the form of new spaces of production, and how the relationships between people and tools, people and capital, and people and authorities need to be remodeled to provide the conditions for radical innovations in the development of novel socio-technical configurations. Peter is an industrial engineer by training (PhD from ETH Zurich, 1999). He worked in factory automation, attaching robots and automatic tool-changers to CNC milling machines, before pursuing his career as a business consultant in Switzerland, and later as a research manager in knowledge technologies and knowledge management at the University of Aberdeen, Scotland, UK, and as a senior project manager at Waag Society in Amsterdam. He is currently a Research Professor on the Revolution in Manufacturing at Rotterdam University of Applied Sciences, The Netherlands.

CECILIA RASPANTI

is a fashion and textile designer. She was born in the Netherlands and grew up in Italy, where she studied fashion design and knitwear at the Polimoda Fashion Institute in Florence. At the beginning of 2013 she started visiting the Fablab Amsterdam, where she came into contact with digital fabrication, and started researching the application of technology and digital fabrication to textile manipulation. Since she started her research project in Fablab Amsterdam in 2013, she has started her own label, Ceciilya, and her main interest consists in creating digital translations of old craft techniques to create new, fashionable materials. Always inspired by nature, her latest collection was based on invertebrates and small living organisms, in which decorative patterns become structural elements of the garments. She currently works at Waag Society on projects related to her field of expertise, and is involved in teaching at the Fabschool kids program.

ALEX SCHAUB

works as researcher on open design at Waag Society. Alex has worked at Waag Society since 2005, initially as head of the technical team and later as manager of Fablab Amsterdam. Alex graduated with a BA in music and in art in 1999, and successfully completed his MA in music in 2001 at the Koninklijk Conservatorium in The Hague. Besides his work as technician and manager, he has been involved in many art projects, such as Sonic Kitchen, an immersive performance that integrated food, sound and image.

HELOISA NEVES

is executive director of Fab Lab Brazil Network (an institution that helps in the implementation of the Fab Lab network in Brazil) and creator of We Fab (a company that connects the Maker movement strategies with the educational, cultural and entrepreneurial world. It currently has two active programs: Fab Educação and Maker Innovation). She is also a partner of Garagem Fab Lab in São Paulo and Professor on the engineering course at Insper, where she also helps with the Insper Fab Lab. She graduated from Fab Academy Barcelona and in the past has collaborated with Fablab Sevilla. She recently translated and republished the book Fab Lab: A Vanguarda da Nova Revolução Industrial with Fabien Eychenne, and wrote a chapter for the book Yes, we are open! – Fabricação Digital, Tecnologías y Cultura Libre by Jose Perez de Lama. She is currently finishing her PhD, entitled 'Maker Innovation – from Open Design and Fab Labs to the strategies inspired by the Maker movement'.

Further Reading

Makers

Anderson, C., 2012. *Makers: The New Industrial Revolution*. Crown Business.

Charny, D., 2011. *Power of Making: The Importance of Being Skilled*. London: V & A Publishing.

Doctorow, C., 2009. *Makers*. London: HarperCollins Publishers.

Gauntlett, D., 2011. *Making is Connecting*. 1st edition, Polity.

Hatch, M., 2014. *The Maker Movement Manifesto. Rules for Innovation in The New World of Crafters, Hackers, and Tinkerers*. New York: McGraw-Hill Education.

Lang, D., 2013. *Zero to Maker: Learn (Just Enough) to Make (Just About) Anything*. Maker Media, Inc.

Osborn, S., 2013. *Makers at Work: Folks Reinventing the World One Object or Idea at a Time*. Apress.

Parks, B., 2005. *Makers: All Kinds of People Making Amazing Things In Garages, Basements, and Backyards*. Sebastopol, CA: O'Reilly Media.

Sennett, P.R., 2009. *The Craftsman*. Yale University Press.

Thomas, A., 2014. *Making Makers: Kids, Tools, and the Future of Innovation*. 1 edition, Maker Media, Inc.

Fab Labs and Makerspaces

Eychenne, F., 2012. *Fab Lab: L'avant-garde de la nouvelle révolution industrielle*. Limoges: FYP Editions.

Burke, J.J., 2014. *Makerspaces: a Practical Guide for Librarians*. Lanham: Rowman & Littlefield Publishers.

Kemp, A., 2013. *The Makerspace Workbench: Tools, Technologies, and Techniques for Making*. Make Books.

Gershenfeld, N., 2005. *FAB: The Coming Revolution on Your Desktop – From Personal Computers to Personal Fabrication*. Basic Books.

Menichinelli, M., 2013. '10 things to do when starting a FabLab'. *openp2pdesign. org*. Available at: http://www. openp2pdesign.org/2013/spaces/10-things-to-do-when-starting-a-fablab/

Menichinelli, M., 2013. 'How to Build a FabLab'. *openp2pdesign.org*. Available at: http://www.openp2pdesign.org/2013/spaces/how-to-build-a-fablab/

Menichinelli, M., 2013. 'What is a FabLab?' *openp2pdesign.org*. Available at: http://www.openp2pdesign.org/2013/spaces/what-is-a-fablab/

Troxler, P., 2011. Libraries of the Peer Production Era. In *Open Design Now: Why Design Cannot Remain Exclusive*. Amsterdam: BIS Publishers.

Walter-Herrmann, J., 2013. *FabLab: Of Machines, Makers and inventors*. Bielefeld: transcript.

Digital Fabrication

Dunn, N., 2012. *Digital Fabrication in Architecture*. London: Laurence King Publishing.

Labaco, R. ed., 2013. *Out of Hand. Materializing the Postdigital*. Black Dog Pub Ltd.

France, A.K., 2013. *Make: 3D Printing: The Essential Guide to 3D Printers*. 1st edition, Maker Media, Inc.

Lipson, H. & Kurman, M., 2013. *Fabricated: The New World of 3D Printing*. Indianapolis, Indiana: Wiley.

Warnier, C. et al., 2014. *Printing Things: Visions and Essentials for 3D Printing*. Berlin: Gestalten.

Working with hardware

Banzi, M., 2009. *Getting Started with Arduino*. Sebastopol: O'Reilly Media.

Borenstein, G., 2012. *Making Things See: 3D Vision with Kinect, Processing, Arduino, and MakerBot*. Sebastopol: Maker Media, Inc.

Karvinen, K., 2014. *Getting Started with Sensors: Measure the World with Electronics, Arduino, and Raspberry Pi*, Sebastopol: Maker Media, Inc.

Igoe, T., 2011. *Making Things Talk: Using Sensors, Networks, and Arduino to See, Hear, and Feel Your World*. 2nd edition, Beijing: Maker Media, Inc.

Jepson, B., 2012. *Learn to Solder: Tools and Techniques for Assembling Electronics*. Sebastopol: Maker Media, Inc.

Platt, C., 2009. *Make: Electronics*. Sebastopol: Make Maker Media, Inc.

Platt, C., 2012. *Encyclopedia of Electronic Components Volume 1: Resistors, Capacitors, Inductors, Switches, Encoders, Relays, Transistors*. Sebastopol: Maker Media, Inc.

Platt, C., 2014. *Make: More Electronics: Journey Deep Into the World of Logic Chips, Amplifiers, Sensors, and Randomicity*. Sebastopol: Maker Media, Inc.

Richardson, M., 2012. *Getting Started with Raspberry Pi*. 1st edition, Sebastopol: Maker Media, Inc.

Valtokari, V., 2014. *Make: Sensors: A Hands-On Primer for Monitoring the Real World with Arduino and Raspberry Pi*. 1st edition, Sebastopol: Maker Media, Inc.

Working with software

Downey, A.B., 2012. *Think Python*. 1st edition, Sebastopol: O'Reilly Media. Available at: http://www.greenteapress.com/thinkpython/.

Lutz, M., 2013. *Learning Python, 5th Edition*. 5th edition, Sebastopol: O'Reilly Media.

Reas, C. & Fry, B., 2010. *Getting Started with Processing*. 1st edition, Beijing ; Sebastopol, CA: Maker Media, Inc.

Reas, C. & Fry, B., 2014. *Processing: A Programming Handbook for Visual Designers and Artists*. 2nd edition, Cambridge, Massachusetts: The MIT Press.

Design resources

Ashby, M.F., 2010. *Materials Selection in Mechanical Design, Fourth Edition*. 4th edition, Burlington, MA: Butterworth-Heinemann.

Griffin, M., 2015. *Design and Modeling for 3D Printing*. Sebastopol: Maker Media, Inc.

Martin, B., 2012. *Universal Methods of Design: 100 Ways to Research Complex Problems, Develop Innovative Ideas, and Design Effective Solutions*. Beverly, MA: Rockport Publishers.

Roberts, D., 2010. *Making Things Move. DIY Mechanisms for Inventors, Hobbyists, and Artists*. New York: McGraw-Hill/TAB Electronics.

Credits

CHAPTER 01

Infographic:

Content: M. Menichinelli
Infographic: Matteo Astolfi

Pictures:

p. 17: Shir Shpatz
p. 18: Fabio Lafauci
p. 23: Fabio Lafauci
p. 27: Camille Bosqué
p. 29: Camille Bosqué
p. 30: Camille Bosqué

CHAPTER 02

Infographic:

Content: M. Menichinelli
Infographic: Matteo Astolfi

Pictures:

p. 36: Juan Peña / SparkFun Electronics
p. 37: From above: SnootLab, David Mellis, Juan Peña / SparkFun Electronics
p. 38: Aalto Fablab
p. 39: From above: Aalto Fablab, Aalto Fablab, Massimo Menichinelli, Aalto Fablab
p. 40: Aalto Fablab
p. 41: From above: Aalto Fablab, Massimo Menichinelli, Massimo Menichinelli
p. 42–43: Aalto Fablab
p. 44: Formlabs
p. 45: Luciano Betoldi
p. 46: From above: Luciano Betoldi, Aalto Fablab, Aalto Fablab, Marc Huguet Reyes
p. 47: From above: Evan Amos, Massimo Menichinelli, Massimo Menichinelli, Massimo Menichinelli
p. 48: From above: Massimo Menichinelli, Massimo Menichinelli, Chris Winters
p. 49: Aalto Fablab

CHAPTER 03

Infographic:

Content: Massimo Menichinelli, Peter Troxler based on Arthur Schmitt, nod-A.
Infographic: Matteo Astolfi

Pictures:

p. 63: Alessandro Randi - CODECZOMBIE

CHAPTER 04

Pictures:

p. 69: John Smith
p. 71: Shapeways
p. 78: Aalto Fablab
p. 79: Aalto Fablab
p. 80: Massimo Menichinelli
p. 81: Massimo Menichinelli
p. 82: Matteo Astolfi, Gianluca Balzerano
p. 83: Matteo Astolfi, Gianluca Balzerano

CHAPTER 05

Infographic:

Content: Massimo Menichinelli, adapted from Iso, 2010. ISO 9241-210:2010 - Ergonomics of human-system interaction - Part 210: Human-centred design for interactive systems. Available at: http://www.iso.org/iso/iso_catalogue/catalogue_ics/catalogue_detail_ics.htm?csnumber=52075

Infographic: Matteo Astolfi

Pictures:

p. 91: Ernst Haeckel
p. 92: From above: Ernst Haeckel, Massimo Menichinelli
p.: Ernst Haeckel

p. 94: Cecilia Raspanti
p. 95: Ernst Haeckel
p. 96: Cecilia Raspanti
p. 97: From above left: Ernst Haeckel, Cecilia Raspanti, Jens Dyvik, Cecilia Raspanti
pp. 98–101: Cecilia Raspanti
p. 102: Tomek Dersu Aaron Whitfield
p. 103: Tomek Dersu Aaron Whitfield
pp. 104–105: Cecilia Raspanti
p. 106: From above: Alex Schaub (model Josephine Keuter), Cecilia Raspanti
pp. 107–108: Alex Schaub (model Josephine Keuter)
p. 109: Cecilia Raspanti

CHAPTER 06

Infographic:

Content+Infographic: Matteo Astolfi

Pictures:

pp. 114–116: Alex Schaub / Carsten Lemme / Alice Mela / Thomas van der Werff
p. 117: From above: first row: Alex Schaub / Carsten Lemme / Alice Mela / Thomas van der Werff. Second row: Aurélie Ghalim. Last row: Alex Schaub / Carsten Lemme / Alice Mela / Thomas van der Werff
pp. 118–125: atFAB (Anne Filson / Gary Rohrbacher)
pp. 126–130: Pietro Leoni
p. 131: Aalto Fablab
pp. 132–134: Paula Studio / Antonio Gagliardi
p. 135: From above: first row: Paula Studio / Antonio Gagliardi. Second row: Paula Studio / Antonio Gagliardi, Picturepest
pp. 136–139: Matthew Keeter
pp. 140–143: Begle Moritz
pp. 144–149: Chirag Rangholia / Aldo Sollazzo
p. 150: Alex Schaub / Angelo Chiacchio /

Robert Nelk / Pepijn Fens

p. 151: Angelo Chiacchio

pp. 152–153: Alex Schaub / Angelo Chiacchio / Robert Nelk / Pepijn Fens

pp. 154–161: IAAC / CBA – MIT / Fab Lab network. Full team credits: http://www. fablabhouse.com/team-2/

pp. 162–165: Luciano Betoldi

pp. 166–171: Jens Dyvik

pp. 172–175: Pietro Leoni

pp. 176–179: Massimo Di Filippo / Jessica Maullu

pp. 180–185: Subalterno1 / Tecnificio

pp. 186–189: Teja Philipp / Jennifer Heier / Philipp Engel

pp. 190–193: Joel Gibbard

pp. 194–195: Open Source Beehives

p. 196: From above: first row: John Rees / Jonathan Minchin / Ferran Masip Valls. Second row: John Rees / Jonathan Minchin / Ferran Masip Valls, Ferran Masip Valls. Third row: Open Source Beehives

p. 197: From above: first row: Open Source Beehives. Second row: Ferran Masip

p. 198: Open Source Beehives

p. 199: From above: first row: Open Source Beehives. Second row: John Rees / Jonathan Minchin / Ferran Masip Valls, John Rees / Jonathan Minchin / Ferran Masip Valls, Fab Lab Barcelona / Institute Of Advanced Architecture Of Catalonia (IAAC) / Media Interactive Design In Collaboration With Hangar / Goteo (Tomas Diez, Alex Posada, Guillem Cam-Prodon, M.A. Heras, Alexandre Dubor, Leonardo Arrata, Xavier Vinaixa, Gabriel Bello-Diaz)

pp. 200–205: Gerard Rubio Arias

pp. 206–211: Flowers-INRIA (Pierre Yves Oudeyer, Matthieu Lapeyre, Pierre Rouanet, Jonathan Grizou, Steve Nguyen, Alexandre Le Falher, Fabien Depraetre)

pp. 212–215: Primo (Matteo Loglio, Filippo Yacob)

pp. 216–221: Enrico Bassi / Maurizio Mion / Andrea Patrucco / Gualtiero Tumolo

pp. 222–225: Fab Lab Barcelona / Institute Of Advanced Architecture Of Catalonia (IAAC) / Media Interactive Design In Collaboration With Hangar / Goteo (Tomas Diez, Alex Posada, Guillem Cam-Prodon, M.A. Heras, Alexandre Dubor, Leonardo Arrata, Xavier Vinaixa, Gabriel Bello-Diaz)

pp. 226–227: Gosuke Sugiyama @ FAB9

pp. 228–233: Ultimaker (Martijn Elserman, Erik de Bruijn, Siert Wijnia)

pp. 234–237: Studio dLux (Denis Fuzii)

pp. 238–240: Afate Gnikou / WoeLab.

p. 241: Daniel Hayduk

p. 242: Cesare Griffa / Massimiliano Manno / Denise Giordana / Federico Borello

pp. 243–245: Cesare Griffa / Matteo Amela

pp. 246–247: Cesare Griffa / Massimiliano Manno / Denise Giordana / Federico Borello

CHAPTER 07

Illustrations:

p. 252: Paul Bizcarguenaga

CHAPTER 08

Infographic:

Content: The Fab Foundation, Labs | FabLabs. *FabLabs.io - The Fab Lab Network.* Available at: https://www.fablabs.io/labs
Infographic: Matteo Astolfi

Illustrations:

pp. 262–263: Paul Bizcarguenaga

Icons:

"Fused Deposition Modelling" by Ollie Taylor
"Handwork" by Created by Scott Lewi
"Molding and Casting" by Jasmin Rae Friedrich
"Laser Cutting" by Nick Green

"Electronic Components" by Kenneth Appiah
"Microcontrollers" by David Waschbüsch
"Sensors" by TagTeam Studio
"Oscilloscope" by Aleksandrs
"Multimeter" by Sergey Krivoy
"Book" icon by Derrick Snider
"Taj Mahal" icon by Anna Weiss
"Globe" by David Vickhoff
"Teacher" by Yazmin Alanis
"Tile" by Samu Parra
All the icons above from The Noun Project

Acknowledgments

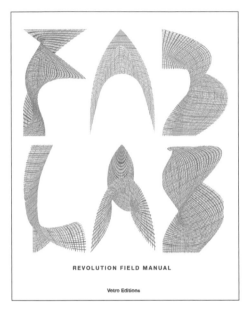

REVOLUTION FIELD MANUAL

Vetro Editions

Alternative book cover proposal by
Loes Verstatten, 75B Rotterdam

First and foremost, I would like to thank all the authors, most of whom are active participants of the global Fab Lab community, for sharing their time, knowledge and experience in contributing content to this book. In addition, thanks are due to all the other members of the community, for their collaborations and discussions on the topics relevant to the community over the years: the strength and prominence of such activities is a sign of the health of the community, which never ceases to discuss its future collaboratively. Most of the contents of this book stem from these collaborations and discussions.

Furthermore, I would like to thank my wife Laura Mata García and my parents, Ornella Marchini and Mario Menichinelli, for supporting me during my years of research on the topics of the book.

Massimo Menichinelli

Vetro Editions would like to thank: Céline Remecido and Michel Chanaud at Pyramyd Editions, who made possible the development of this project, and inspired us to pursue it; Markus Sebastian Braun at Niggli for publishing the English version; the guys at Fab Lab Berlin who welcomed us when we didn't even know what a 3D printer did; Dario Buzzini who directed us to the capable supervision of the author, Massimo Menichinelli; the author himself, for his incredible patience and great abilities; John Z. Komurki and Matteo Astolfi for their constant support; the organization team at FAB10 Barcelona, Paz Diman, Alessandro Randi and 75B Studio Rotterdam; and everyone else who helped us with this project.